PROGRESS IN BIOMASS AND BIOENERGY RESEARCH

PROGRESS IN BIOMASS AND BIOENERGY RESEARCH

STEVEN F. WARNMER
EDITOR

Nova Science Publishers, Inc.
New York

NOTICE TO THE READER

The Publisher has taken reasonable care in the preparation of this book, but makes no expressed or implied warranty of any kind and assumes no responsibility for any errors or omissions. No liability is assumed for incidental or consequential damages in connection with or arising out of information contained in this book. The Publisher shall not be liable for any special, consequential, or exemplary damages resulting, in whole or in part, from the readers' use of, or reliance upon, this material.

Independent verification should be sought for any data, advice or recommendations contained in this book. In addition, no responsibility is assumed by the publisher for any injury and/or damage to persons or property arising from any methods, products, instructions, ideas or otherwise contained in this publication.

This publication is designed to provide accurate and authoritative information with regard to the subject matter cover herein. It is sold with the clear understanding that the Publisher is not engaged in rendering legal or any other professional services. If legal, medical or any other expert assistance is required, the services of a competent person should be sought. FROM A DECLARATION OF PARTICIPANTS JOINTLY ADOPTED BY A COMMITTEE OF THE AMERICAN BAR ASSOCIATION AND A COMMITTEE OF PUBLISHERS.

Library of Congress Cataloging-in-Publication Data
Progress in biomass and bioenergy research / Steven F. Warnmer (editor).
        p. cm.
Includes bibliographical references and index.
ISBN 13 978-1-60021-328-1
ISBN 10 1-60021-328-6
1. Biomass--Research. 2. Biomass energy--Research. I. Warnmer, Steven F.
TP360.P768   2006
662'.88--dc22   2006020634

Published by Nova Science Publishers, Inc. ✤*New York*

CONTENTS

PREFACE

Like coal and petroleum, biomass is a form of stored solar energy. The energy of the sun is "captured" through the process of photosynthesis in growing plants. Like all methods used to generate energy, the combustion of biomass generates pollution as a by-product. One advantage of biofuel in comparison to most other fuel types is that the energy within the biomass can be stored for an indefinite amount of time without any danger.

Agricultural products specifically grown for use as biofuels include corn and soybeans, primarily in the United States, as well as flaxseed and rapeseed, primarily in Europe, and hemp is a growing crop around the world except for in America. Waste from industry, agriculture, forestry, and households can also be used to produce bioenergy; examples include straw, lumber, manure, sewage, garbage and food leftovers. Biomass used as fuel often consists of underutilized types, like chaff and animal waste. Much research is currently in progress into the utilization of microalgae as an energy source, with applications being developed for biodiesel, ethanol, methanol, methane, and even hydrogen. On the rise is use of hemp, although current politics restrains it. This new book presents the latest leading edge research in a field set to explode with growth.

Chapter 1- Plants represent a natural chemical and polymer factory and food plant. Biorefineries combines necessary technologies between biogenic raw material and intermediates and final products. The paper present two strategies for producing of polymeric materials, firstly the utilization of the pre-determined natural macromolecular structure and secondly the using of biogenic building blocks. The first step is the fractionation technology from green biomass for producing of fiber-rich press cake and a nutrient rich-green juice. The main focus is directed on products, such as proteins, polylactic acid, cellulose and levulinic acid- sequence products and their application as well as their market.

Chapter 2- In this chapter a set joint of experimental techniques for assessing biomass combustion devices is presented. Small scale energy converters such as chimneys, boilers, stoves, etc, producing heat and/or hot water by combustion of biomass (wood, pellets, briquettes, etc.) are especially suited to domestic purposes. However, in regular commercial combustion conditions, this kind of use still has some disadvantages: besides the fact that some emissions (volatile organic carbons, carbon monoxide or NOx) may still be high, it is difficult to compare the quality and performance of equipment working in very different combustion conditions.

Due to their relatively low cost and the complexity of combustion in such devices, modelling by numerical analysis is seldom attempted. Controlling operational factors are usually designed and regulated based on the manufacturer's experience or on handbook values. In order to protect customers, and to assure compliance with minimum requirements for energy performance and maximum limits on pollutant emissions, several national and international regulations have been developed in recent years. Experimental analysis of these devices is a key technique for control and improvement.

Chapter 3- Mitigation of and adaptation to climate change belong to the most pressing global challenges for the 21st century. Major mitigation options include improved energy efficiency, shifting towards less carbon-intensive fossil fuels, increased use of energy sources with near-zero emissions, such as renewables and nuclear, CO_2 capture and permanent storage (CCS), and carbon sequestration by protection and enhancement of biological absorption capacity in forests and soils.

Bioenergy is one of several energy sources which could provide society with energy services with near-zero emissions. Bioenergy has a unique feature, however, which distinguishes it from other low-emitting energy supply options, such as solar, wind, nuclear, and clean fossil energy technologies. Bioenergy conversion could be integrated with a process which separates carbon. If the biomass feedstock is sustainably produced and the separated carbon is subsequently isolated from the atmosphere for a very long time the entire process becomes a continuous carbon sink – in other words such technologies yield negative CO_2 emissions. Negative emission biomass technologies can be centralised or distributed; Centralised negative emission biomass technologies, biomass energy with CO_2 capture and storage (BECS), build on the conversion of biomass into energy carriers in centralised conversion plants integrated with CO_2 capture. The captured CO_2 is subsequently transported and stored in geological formations. Distributed negative emission biomass technologies are based on the production of long-term carbon-sequestering charcoal soil amendment, with or without co-production of biofuels.

In this chapter a BECS implementation scenario study is presented. The study analyses investments in BECS in a pulp and paper mill environment. The investment analysis is carried out within a real options framework taking into account the potential revenue from trading generated emission allowances on a carbon market. Uncertainty is considered in the economic modelling through the use of stochastically correlated price processes of one input price (biomass) and two output prices (electricity and CO_2 emission permits) that are consistent with shadow price trajectories of a large-scale global energy model. The results suggest that BECS can be economically feasible within approximately 40 years.

The chapter also discusses Research and Development needs for better understanding of the future overall potential of negative emission biomass technology implementation.

Chapter 4- There is a steady and continuing interest in biomass gasification in both the developed countries and developing countries. While the advanced countries are interested primarily from considerations of reduced emissions and waste utilisation, the developing countries look at biomass gasification as a means to augment commercial energy like electricity, diesel, fuel oil etc.

India, a tropical country with a vast geographical area is richly endowed with renewable energy sources like solar, wind, biomass which can play a crucial role in meeting end use energy needs in a decentralised manner. One of the major goals of the ninth and tenth five year plan is strengthening of infrastructure (energy, transport, communication, irrigation) in order to support the growth process on a sustainable basis. It is usually the tendency of the developing countries to equate development with economic growth and to further equate economic growth with energy consumption especially electricity. India being a developing country has also given due emphasis on strengthening its energy position accordingly. Moreover threat from Green House Gasses (GHG) also has caused worldwide concern. In India electric power generation is the largest source of GHG emissions. It accounts for 48% of carbon emitted. These concerns point towards more rational energy use strategies. The renewable and recycling process makes biomass possible to generate power without adding to air emissions.

Biomass (firewood, agricultural residue, and dung) is one of the main fuels in India, particularly in the energy-starved rural sector. The biomass power potential in India was 16,000 MW (excluding co-generation), but the achievement in this respect is negligible (Installed capacity - 630 MW Project under implementation - 630 MW, as on March 2005). It brings out the fact that much of the potential of biomass gasification is still unexplored. Globally, India is in the fourth position in generating power through biomass and with a huge potential, is poised to become a world leader in utilization of biomass.

According to the Planning Commission of India, in its Tenth Five Year Plan, announced that 26.10 per cent of the Indian populations are below the poverty line and mostly belongs to rural areas. The inequitable distribution has been evident from the fact that although 70% of India's population lives in the rural areas, only 29% of rural households have electricity supply as against 92% of urban households. Of the half a million or so villages in India, about 3, 10,000 villages have been declared to be electrified and 80,000 more villages remain completely un-electrified. There are a number of constraints to supply power to remote rural area such as small human settlements, geographically dispersed villages, seasonally of loads etc. In the absence of adequate network and hence supply of power to remote rural areas the household depend largely on primary energy sources like kerosene and diesel for lighting. No commercial investments in micro enterprises can therefore be made by either individuals or companies without installing diesel generators which have a very high generating cost. Biomass gasifier is a leading option in that respect. Besides, the supply of power to remote rural areas from the centralised grid is not competitive than a modern biomass gasification based decentralised power plant. Estimate from an Indian village shows that modest 50 kW of installed capacity per village will lead to total saving of 52000 million Rs (Rs 5200 Crore / 1100 million US $) in power plant investments. In energy terms, the saving in TandD losses will release a generation capacity of 800 MW for profitable sale. Reduced pollution and reduction of CO_2 emissions will be the other advantages of a decentralised renewable energy based system for the rural areas.

The purpose of the present paper is to evaluate the rural electrification programme in India undertaken by the Ministry of Non Conventional Energy Sources (MNES), Government of India, through biomass gasifier power plant. It explores the eradication of poverty that has been made possible by introducing biomass gasification based power plant in remote rural areas in India. Creation of jobs in the power stations, small-scale business, commerce and industries and also improvement in the quality of life is assessed. The paper concludes with policy options relevance for the other developing countries.

Chapter 5- In this study energy balance and fuel properties of biodiesel has been calculated. Accordingly, the cost of 1 liter of oil is calculated 0.32 € after the income from the seed meal is deduced. Finally, the cost of per unit of biodiesel (1 liter) was calculated as 0.55 €, after deduction of the income provided by the sales of glycerin for use in soap and cosmetic industry.

The energy equivalent of total output was calculated 147605.50 MJ per hectare. The net energy gain (refined oil) was found as 15105.63 MJ per hectare (The net energy ratio 11.031) according to yield and inputs values.

The viscosity values of vegetable oils vary between 27.2 and 53.6 mm^2/s whereas those of vegetable oil methyl esters between 3.59 and 4.63 mm^2/s. The flash point values of vegetable oil methyl esters are highly lower than those of vegetable oils. The flash point values of vegetable oil methyl esters are highly lower than those of vegetable oils. An increase in density from 860 to 885 kg/m^3 for vegetable oil methyl esters or biodiesel increases the viscosity from 3.59 to 4.63 mm^2/s and the increases are highly regular. There is high regression between density and viscosity values vegetable oil methyl esters. The relationships between viscosity and flash point for vegetable oil methyl esters are irregular. An increase in density from 860 to 885 kg/m^3 for vegetable oil methyl esters increases the flash point from 401 to 453 K and the increases are slightly regular.

The LHV values of vegetable oils methyl ester vary between 35.74 and 39.16 MJ/kg.

Chapter 6- Instability and increases in prices of petroleum-based fuels, gradual depletion of world petroleum reserves and increases in environmental pollution caused by exhaust emissions speed up research on renewable alternative fuels.

Vegetable oils have been considered as renewable alternative fuels in compression ignition engines for a long time. However, they have not been widely used as fuels in the engines due to some technical and economical drawbacks. Some properties of vegetable oils such as high viscosity, lower volatility and lower heat content result in technical problems in direct using of vegetable oils in short and long term applications. From economical point of view, the main problem is that vegetable oils have been more expensive than petroleum Diesel fuel.

There are various ongoing studies on solving these problems to be able to use vegetable oils in Diesel engines. Different methods such as preheating oils, blending or dilution with other fuels, thermal cracking/pyrolysis and transesterification have been developed. Among these techniques, transesterification appears to be the most promising one. It is a chemical process converting vegetable oils to alcohol ester of oil named as biodiesel. In general, biodiesel-Diesel fuel No.2 blend can be used as a fuel in Diesel engines without modification. Specifications of biodiesel mainly depend on oil, transesterification process, type and amount of alcohol, type and amount of catalysis, reaction time and temperature.

Biodiesel can be produced from different kinds of vegetable oils. Since prices of edible vegetable oils are higher than that of Diesel fuel No. 2, waste vegetable oils and non-edible crude vegetable oils are mostly preferred as potential low priced biodiesel sources. It is also possible to use soapstock, a by-product of edible oil production, for cheap biodiesel production.

In this study, various biodiesels were produced from raw vegetable oils (rapeseed oil, soybean oil, cotton seed oil, palm oil and tobacco seed oil), waste sunflower vegetable oils and hazelnut oil soap stock-waste sunflower vegetable oil, and their specifications were compared with each other. The biodiesel (20% in volume) - Diesel fuel No.2 (80% in volume) blends were tested in a four cycle, four cylinder turbocharged indirect injection Diesel engine. The effects of biodiesel addition to Diesel fuel No.2 on the performance and emissions of the engine were investigated at full load. Experimental results showed that the biodiesels can be partially substituted for Diesel fuel No.2 at most operating conditions in terms of performance parameters and emissions without any engine modification and preheating of the blends.

Chapter 7- Lignin, obtained through steam explosion from straw, was completely characterized via elemental analysis, gel permeation chromatography, ultraviolet and infrared spectroscopy, ^{13}C and ^{1}H nuclear magnetic resonance spectrometry.

Lignin powder was used for the preparation of blends with low-density polyethylene (LDPE), linear low-density polyethylene (LLDPE), high-density polyethylene (HDPE) and atactic polystyrene (PS).

The obtained blends are processable through the conventional techniques used for thermoplastics; the modulus slightly increases for most lignin-polymer blends, while the tensile stress and elongation reduce. Moreover, lignin acts as a stabilizer against the UV radiation for PS, LDPE and LLDPE.

Polyurethanes were obtained treating steam exploded lignin from straw with 4,4'-methylenebis(phenylisocyanate), 4,4'-methylenebis(phenylisocyanate) – ethandiol, and poly(1,4-butandiol)tolylene-2,4-diisocyanate terminated. The obtained materials were characterized by using gel permeation chromatography, infrared spectroscopy and scanning electron microscopy. Differential scanning calorimetry analysis showed a T_g at -6 °C, assigned to the glass transition of the poly(1,4-butandiol) chains. The presence of ethylene glycol reduced the yields of the polyurethanes. The use of the prepolymer gave the best results in polyurethanes formation. Steam exploded lignin was used as starting material in the synthesis of polyesters. Lignin was treated with dodecanoyl dichloride. The products were characterized by using gel permeation chromatography, infrared spectroscopy, ^{13}C and ^{1}H nuclear magnetic resonance spectrometry, and scanning electron microscopy.

In: Progress in Biomass and Bioenergy Research
Editor: Steven F. Warnmer, pp. 1-32

ISBN: 1-60021-328-6
© 2007 Nova Science Publishers, Inc.

Chapter 1

BIOBASED POLYMERS BY CHEMICAL VALORIZATION OF BIOMASS COMPONENTS

B. Kamm[*1], M. Kamm[2], I. Scherze[3], G. Muschiolik[3] and U. Bindrich[4]*

[1] Research Institute of Bioactive Polymer
Systems e.V. Research Center Teltow-Seehof
Kantstraße 55 D-14513 Teltow, Germany
[2] Biorefinery.de GmbH, Potsdam, Germany
[3] FS-University Jena, Department of Food Technology
[4] DIL (Deutsches Institut für Lebensmitteltechnik) e.V., Quakenbrück

ABSTRACT

Plants represent a natural chemical and polymer factory and food plant. Biorefineries combines necessary technologies between biogenic raw material and intermediates and final products. The paper present two strategies for producing of polymeric materials, firstly the utilization of the pre-determined natural macromolecular structure and secondly the using of biogenic building blocks. The first step is the fractionation technology from green biomass for producing of fiber-rich press cake and a nutrient rich-green juice. The main focus is directed on products, such as proteins, polylactic acid, cellulose and levulinic acid- sequence products and their application as well as their market.

Keywords: green biomass, biorefinery, proteins, poly(lactic acid), cellulose, levulinic acid

---

*   Corresponding author: Research Institute of Bioactive Polymer Systems e.V. Research Center Teltow-Seehof Kantstraße 55 D-14513 Teltow, Germany; e-mail: kamm@biopos.de; Tel.:0049-3328-332210; Fax: 0049-3328-332211

ABBREVIATIONS

BR	Biorefinery
BJ	Brown Juice
DM	Dry matter
dt	deciton, decimal metric tonne (1dt = 100 kg)
e.g.	(for example)
GBR	Green biorefinery
GJ	Green Juice
GLNC	Green leaf nutrient concentrate
ha	hectare (10.000 m^2 =100m x 100m)
kg	Kilogramme (1kg = 1000 gramme)
LNC	Leaf nutrient concentrate' (LNC)
LP	Leaf protein
LPC	Leaf proteine concentrate
PC	Press cake
pound (eng)	1 pound (germ) = 1.1023 pound (eng)
SJ	Silage Juice
SPC	Silage press cake
t	ton, decimal metric tons (1t = 1000 kg)
t/y	ton(s) per year

INTRODUCTION

Sustainable economical growth requires safe resources of raw materials for the industrial production. Today's most frequently used industrial raw material, petroleum, is neither sustainable, because limited, nor environmentally friendly. While the economy of energy can be based on various alternative raw materials, such as wind, sun, water, biomass, as well as nuclear fission and fusion, the economy of substances is fundamentally depending on biomass, in particular biomass of plants. Special requirements are placed to both, the substantial converting industry as well as research and development regarding the efficiency of the product line as well as sustainability. "The development of biorefineries represents the key for the access to an integrated production of food, feed, chemicals, materials, goods, and fuels of the future" (National Research Council, 2000).

Many of the currently used industrially made biobased products are results of a directly physical or chemical treatment and processing of biomass, such as cellulose, starch, oil, protein, lignin and terpenes. By biotechnological processes and methods feedstock chemicals are produced such as ethanol, butanol, acetone, lactic acid and itaconic acid as well as amino acids, e.g. glutamic acid, lysine, tryptophane. On the other side, currently only 6 billion tons of the yearly by photosynthesis produced 170 billion tons biomass are used; in addition, only 3 percent of these in the non-food area, such as chemistry (Zoebelin, 2001). The today's product lines in the chemical industry produce a few basic chemicals from petrochemical raw material which represent the basis for the synthesis of a wide product palette for nearly all life areas.

The development of comparable biorefineries – however not in the sense of copy – is necessary to produce a broad variety of biobased products in an efficient construction set system. Each biorefinery refines and converts its corresponding biological raw materials into a multitude of valuable products. The product palette of a biorefinery includes not only such products also producable in a petroleum refinery, but in particular such products, which are not accessible in petroleum refineries. Therefore, it is necessary to develop new biorefinery basis technologies, such as (1) (LCF)-Lignocellulosic Feddstock Biorefinery, (LCF)-pre-treatment and effective separation into lignin, cellulose and hemicellulose, (2) further development of thermal, chemical and mechanical processes, such as extractive methods, gasification (syngas) and liquefaction of biomass, (3) further development of biological processes (biosynthesis, bacteria for degradation of starch and cellulose, etc, (4) combination of substantial conversions, such as biotechnological and chemical processes; (5) corn-biorefinery-concepts, (6) green biorefinery-concepts (7) promotion of research and development into phase III-biorefinery: feedstock-mix + process-mix ➜ product-mix, (Kamm and Kamm, 2004). (Figure 1).

Therefore well-known technologies and methods have to be applied in a combinatory way.

From today's point of view there are two principle ways for the utilization of the synthesis power of the nature for the application area of degradable polymers based on biogenic raw materials:

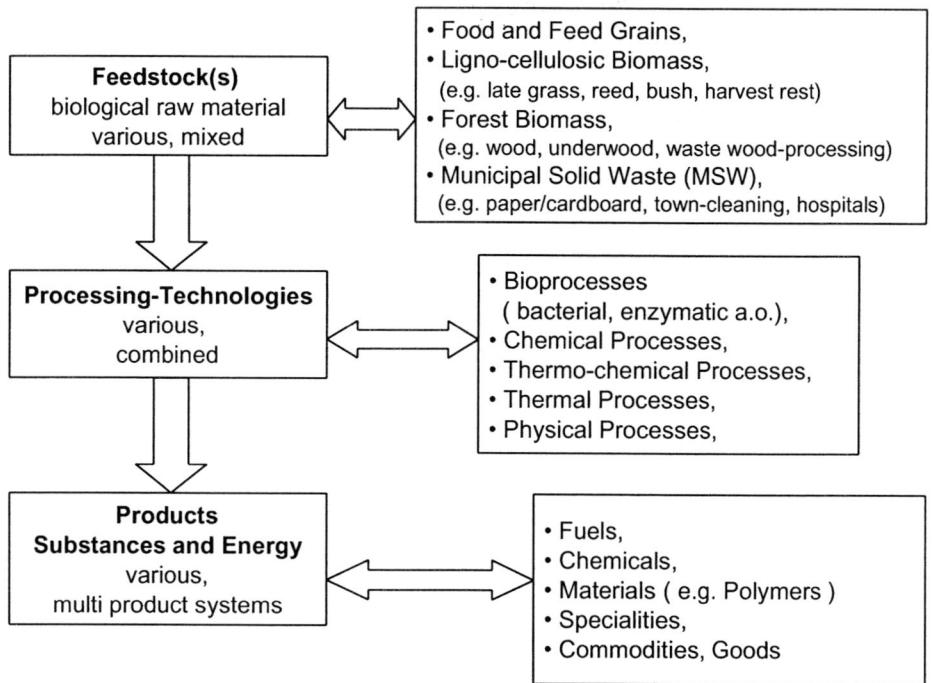

Figure 1. Basic principles of a biorefinery (Type III biorefinery).

Strategy I. Polymeric Materials from Biogenic Macromolecular Structures

First way is the utilization of the pre-determined natural macromolecular structure under extensive maintenance of this specific structure, if necessary also of parts of the complex plant morphology and direct modification of properties for particular applications.

Included are the following classes of substances: (1) nucleic acids, (2) proteins, (3) polysaccharides, (a) poly(α)-glucoses — starch and dextrins, (b) poly(ß)-glucoses — cellulose, lignocellulose, (c) xylane — hemicellulose, (d) poly(galactoses) — pectins, (e) poly(mannoses) — alginates, (f) poly(ß-glucosamines) — chitins and (4) poly(hydroxy-fatty acids). These are hetero chain polymers in form of ester -, amid-(peptid-) and/ or glycosid-structures, which are hydrolytically degradable, that means acid- or base-catalyzed as well as enzymatically (Ebert, 1993).

Penetration of water and thus degradability of polymers can be influenced by means of changing the physical structure, as (plastic) shaping and/or modification of the chemical structure, as increasing or decreasing of hydrophobicity as well as hydrophilicity of the corresponding polymers. In particular for functional applications of renewable raw materials, as fibre composite, starch-resultant products etc. this way is followed (Sixth Symposium on Renewable Resources and Fourth European Symposium on Industrial Crops and Products, 1999).

For applications as basic chemical building blocks these polymers have limits due to their non-uniform structures depending on the respective quality of the nature-built batch.

Strategy II. Polymeric Materials from Biogenic Building Blocks

Main requirement in the construction system of chemistry are uniformly structured compounds, which are converted via tailor-made syntheses into highly processed degradable structures (Verband der Chemischen Industrie, 1994).

This is the guidance of the second principle of the synthesis of degradable structures: Biogenic raw materials can be degraded to well-defined uniform monomer structures by means of biotechnological or chemical methods. These building blocks can then be used for the synthesis of the target compounds. Ideally, the breakdown and build-up of the polymer structures are then combined.

By means of chemical degradation of hexoses- as well as pentoses-containing raw materials well-defined structures such as levulinic acid (γ-oxocarbonic acid), hydroxymethylfurfural (HMF) or furfural are available. Currently applied monomers biotechnologically produced from hexosenic raw materials are (1) α-hydroxycarbonic acid, as lactic acid, malic acid, (2) olefinic carbonic acid, as fumaric acid, itaconic acid (3) polyvalent alcohols, as 2,3-butanediol, 1,3-propanediol, dihydroxyacetone, (4) α-aminocarbonic acid, as glutamic acid and lysine as well as (5) subsequent products such as carnitine (ß-hydroxybetaine) (Kamm, 2004).

2. THE GREEN BIOREFINERY

2.1. Principles

Green biorefinery represents a complex system of ecological technologies for the comprehensive (holistic) substantial and energetic utilization of renewable raw and natural materials in form of green and waste biomass from a targeted sustainable regional land utilization. Such green biomass are for example grass from cultivation of permanent grass land, closure fields, nature preserves or green corps, such as lucerne, clover, immature cereals from extensive land cultivation. Thus, green plants represent a natural chemical factory and food plant. The careful wet fractionation technology is used as first step (primary refinery) to isolate the content-substances in their natural form. Thus, the green crop goods (or humid organic waste goods) are separated into a fiber-rich press cake (PC) and a nutrient-rich green juice (GJ). Beside cellulose and starch, the press cake contains valuable dyes and pigments, crude drugs and other organic substances. The green juice contains proteins, free amino acids, organic acids, dyes, enzymes, hormones, minerals, high-quality crude drugs and other organic substances. By the help of the bio-technology, the eco-technology, the 'soft' and 'green' chemistry, these valuable materials can be isolated in their natural form, or via mild conversion carefully be devoted to an economical utilization (Kamm et al., 2000).

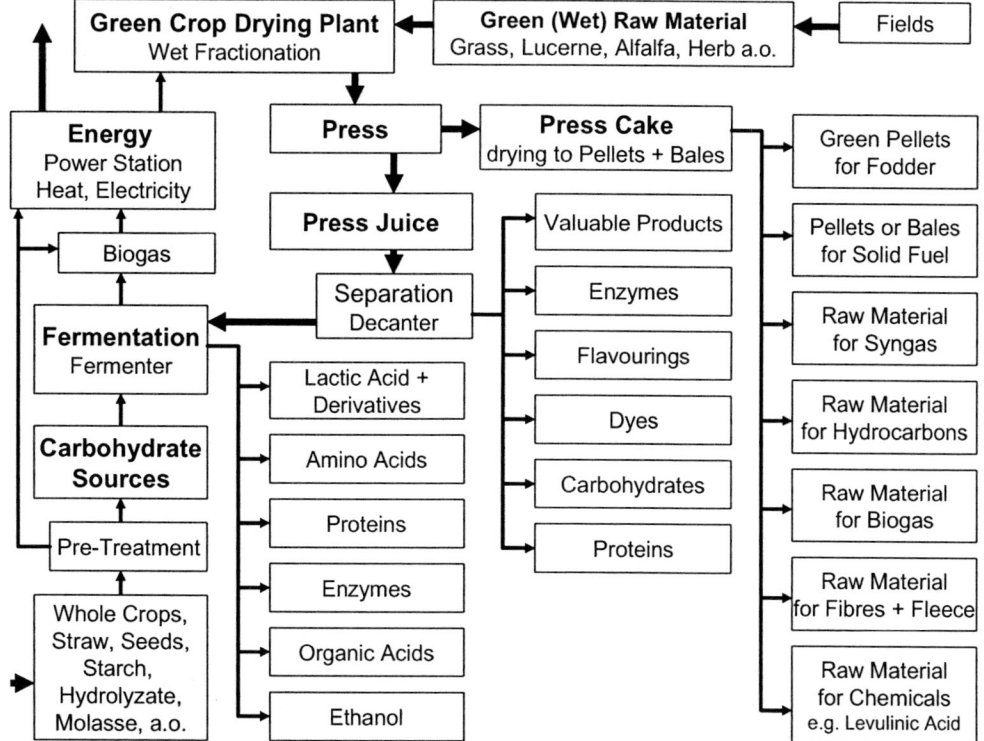

Figure 2. A System Green Biorefinery combined with a green crop drying plant.

Concept of Havelland-Biorefinery, Selbelang, State of Brandenburg, Germany

The activities in the field green biorefinery system grown more and more and developed into an independent line within the large field of biomass technologies. Beside raw material and technology aspects, this system is particularly characterized by the approach of consideration and attention of sustainability criteria (Hector et al.) and incorporation of technologies in regional living and business spaces (sustainable economy, sustainable agriculture, sustainable regional development).

The term 'Green Biorefinery' is on the one hand-side used for model procedures, but on the other hand-side also for a entire program.

To *'refine'* is originally French (raffiner) and means 'something to improve, to purify'. A refinery is per definition a technical facility for the purification, separation and refinement of materials and products. 'Green' in the field of plants means the simultaneousness of high concentration of chlorophyll, nutrients and water, 'Bio' is Greek (bios) and means 'live', something biological and natural. Programmatically, The Green Biorefinery stands for a technology (refinery), formed by the nature imitated (biologically) with the target to be careful, sustainable and ecologic.

The Green Biorefinery pursues the following approaches. The Green Biorefinery represents

- a complex system of ecological technologies
- a model for the study of ecological process management
- finally, an economically self-consisting enterprise and economic entity, respectively.

Therefore, the green biorefinery can be defined as follows.
The green biorefinery is:

- **a complex system of ecological technologies** for the comprehensive material and energetically use and utilization of renewable raw and natural materials in form of green and waste biomass from a targeted sustainable regional land utilization.
- **a model for the study of ecological process management**, that means for the environmental friendly reorientation of the production and energy supply under the premise of sustainability. This model includes the following fields:
  - the supply of raw materials from sustainable, that means environmental and social friendly land utilization,
  - regional sustainable economic procedures based on modern stock-flow management ,
  - the development of a value-material oriented agriculture
  - the step-wise replacement of material and energy management of fossil raw materials by technology transfer,
  - the introduction of ecological technologies and products into market and practice.
- **an economic self-consisting business** of complex technical facilities for the purification, separation and refination of renewable raw materials in form of green and waste biomass together with a self-supporting energy supply on the basis of

renewable raw materials, e.g. bioenergy via biogas, an in-plant material circulation, in particular water and an ecological water and waste treatment.

2.2. Raw Materials

A main raw material is the 'green biomass'.

- This includes the large group of green plant materials: green grasses (meadows, willows, natural resources, extensive willow management), the wild fruit and crops lucerne (table 1) and clover as well as immature cereals and plant shoots. The green plant material contains complex natural and value materials in form of carbohydrates, proteins, fibres, fragrances, dyes, fats, hormones, amino acids, enzymes and others (Pirie, 1971; Carlsson 1989, Carlsson 1997). By primary production of photosynthesis in green plants more than 20 tons of dry matter and 3 tons of protein per ha in temperate climates can be obtained per year (Carlsson 1985).

Table 1. Crude components of Medica sativa L. (lucerne alfalfa)

Medica sativa L. alfalfa, lucerne, green plant					
ingredients	Nitrogen free extractives**	Crude fibre	Crude proteins	Crude fat	Crude ashes ***
components in wt %	6-14,4	3,5-13,4	2,8-7,3	0,5-1,0	1,8
yield /ha* in tons [t]	0,55-1,32	0,32-1,23	0,27-0,67	0,046-0,092	0,165
*Lucerne yield 9,3 t/ha/harvest (DM), DM-dry matter, harvest in july (Robowsky 1998), **Nitrogen free extractives: crude drugs, sugar, dyes and pigments, enzymes, vitamines, free acids a.o., ***Ashes: Ca, P, K, Mn, trace elements Fe, Zn, Cu, (Hagers handbook of pharmaceutical practice, 1972-78)					

- The second large raw material source are the green harvesting residue materials from agricultural cultivated crops. In particular such vegetables, that are harvested with green foliages. This includes e. g. not insignificant amounts sugar beet leaves (sugar beet for sugar industry) (table 2), hemp scrapes and leaves (hemp for fibre production), residue from flax processing, residue of the fresh vegetable production.

Table 2. Yields of sugar beet herbage/foliage per year in the district Havelland

Beta vulgaris/altissima
(sugar beet), yields on
herbage/foliage

area under cultivation*		1225 ha (1998)
yields of sugar beet herbage/foliage**	35-40 t/ha	42.875,0 -49.000,0 t/y
dry matter substance (DM)**	15-18 (ϕ16,5) %/ha	ϕ 16,5%
yields of sugar beet herbage/foliage (DM)	5,77-6,6 t/ha (DM)	7.074,0-8.085,0 t/y
*district Havelland, Germany, (State of Brandenburg Brandenburg, (2003), ** (Fechner and Hertwig, 2003) t/y = tons per year		

- Further potential refinery raw materials are the less normed (standardized) juice-rich waste biomass. These should contain moisture and on the top listed natural and value- materials or also conversion grade. According to coupling effects of material and energetic use, the constitution can strongly vary. Such waste biomass are not standardized goods, but renewable natural waste good that has mainly to be waste managed. This can be residuals of plant production (mixed and ripe harvest residuals), potato juices, hydroxycarboxylic acid-rich wastes, as silage seepage, juices of the canned foods industry, remainder of the sugar industry, remainder of the animal production.

- The 4th large group are less normed (standardized) dried biomass and waste bio-mass. These often contain a high amount of plant cellulose and will therefore be supplied as raw material to press-cake-using production lines. This can be residual straw, hay and all kinds of dried foliage (e.g. maize hay). But also residuals of in-plant waste paper and wood, e.g. for energy production or cardboard production. This group also includes modern concepts of dry crop fractionation, as immature cereals (Coombs and Hall, 1997).

It should be mentioned, that the transitions between raw material types will and should be fluid.

2.3. Primary Technologies

The special feature of the green biorefinery is the wet fractionation or watery-fractionation of green biomass (Figure 3: way A).

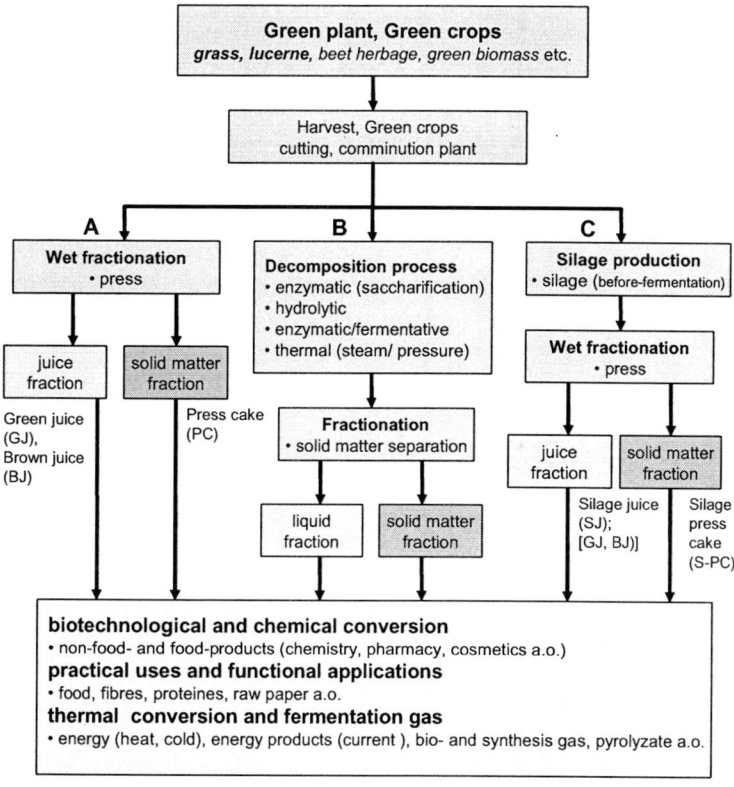

Figure 3. Green Biorefinery — Primary refinery. Methods for fractionation of green crops.

This is also called first fractionation step or primary refinery step. (This includes for example the harvest, fractionation, conservation and storage of the primary fraction). Here, fresh harvest and waste goods are treated. Thus, the plant compounds are mostly unadulterated; however, the green good should in any case immediately be re-worked. This process step, in generally by technical press produced a faser-rich quantity of water-unsoluble solid material, Press cake (PC) and a nutrients-rich Green juice (GJ) or Brown juice (BJ). The wet fractionation based on the soft separation of water soluble and water-unsoluble components of the green biomass.

The silage wet-fractionation is a form of the primary refinery technology (Figure 3, way C). The green goods are conserved by organic acids or fermentation processes before treating them in the following procedure. The treatment of silage green goods have any advantages (decentral raw material preparation, simple and low price conservation and storage, reasonable whole year operation of the processing-step and other more (Kromus et al. 2004). The use of the end products of silage is restricted because silage-chemicals attacks the cell walls and can modify substances.

The so-called breaking up methods are the 3rd category of the primary refinery technology (Figure 3, way B). Breaking up methods consider mainly the humid or dry whole plant. The use procedures are working by enzymatic, fermentative, hydrolytic, chemical, thermal or thermal in combination with press methods. The strength (deepness of operation) of the breaking up methods is differently, and ranges from a low (enzymatic, fermentative) to a high level (chemical, hydrolytic). For every step a classification is needed to check if it belongs to the green biorefinery technology. A high single yield of products can be achieved if the complete plant breaking up methods takes place at the primary refinery step (e.g. by saccharification, increases the total amounts of sugar of the raw charge. But these procedure decrease the level of the utilization and product diversity. Nevertheless breaking up methods have been consider to be usable from technological and economic point of view. This is also the case for the secondary product lines. Green biorefinery-system contradiction can be solved by a further development of new biotechnological breaking up methods.

All primary fractionations following the secondary refinery steps contain processes for substantial and energetical utilization of fractionation products. The kind and number of the secondary fractionation steps are determinate by composition and energy potential of the input green biomass and waste biomass, the status of the technology as well as the market ability of potential products of refinery.

3. PRODUCTS FROM THE PRESS CAKE

The kind and number of products of a green biorefinery is nearly unlimited, if the fractal character of the biosynthesis and biochemistry of the green plant materials is considered (Peitgen and Richter, 1986). The characterization of the plant by a new analysis method usually discovers in addition to the main product innumarable new products. Not all ingredients have been discovered and technologically gained from natural products even from plants with large trade importance as e.g. lucerne (Starke et al. 2000). For the green biorefinery the main products, beside products as well as charge-impurities are interesting. However the ecotechnology and the biotechnology are determinate by the cost-use-efficiency. The economic aspect reduces the diversity of products, also soft technologies are used to reduce the complex molecules of nature materials. Nevertheless the scientific branch of eco-technology tries to develop new methoda, preferring a reducing of strengths (deepness of operation). This can be done by using e.g. biodiversity before molecule modification or by applying low injure degree methods etc. (Moser, 1997).

Following products and group products are possible (technological; after fractionation according to variant A):

From the solid matter fraction (press cake, PC) [see also Fig 2]

- The application of the PC as feeding stuff (silage, bale press food, green pellets (Fechner, 1998)
- The utilization of the PC as source of energy (burning, fermentation gas) (Hertwig and Scholz, 1998)
- The extraction of the dyes of plant (chlorophyll, carotene, xanthophyll) (Schertz, 1938) and application such in the food and candle industry (Judah, 1954) and environmental analytical or after pure refining in the cosmetic, medicine, biochemistry (Wantanabe 1983), electronic, as nematic liquid crystals (Leblanc et al., 1984) and photovoltaic, as organic dyes (Meissner,1997).

Due to structural similarities of chlorophyll and blood hemoglobine one can expect interesting developments in the field of plant dyes and colorants. The resulting fraction will materially and thermally be treated analogue to PC. The suitability (and applicability) as feed depends mainly on the corresponding extraction compounds and has to be tested.

The press cake fraction can be separated analogue to wood raw materials into its main components (Figure 4). On the one side, this green plant press cake fractionation seems from an economical point of view to make not much sense today (due to wood competition). On the other side, there are interesting applications for special vegetable celluloses, hemicelluloses, lignins and monosaccharides. However, short-term possibilities could be the partial fractionation in combination with other applications (paper, cardboard, moulded articles).

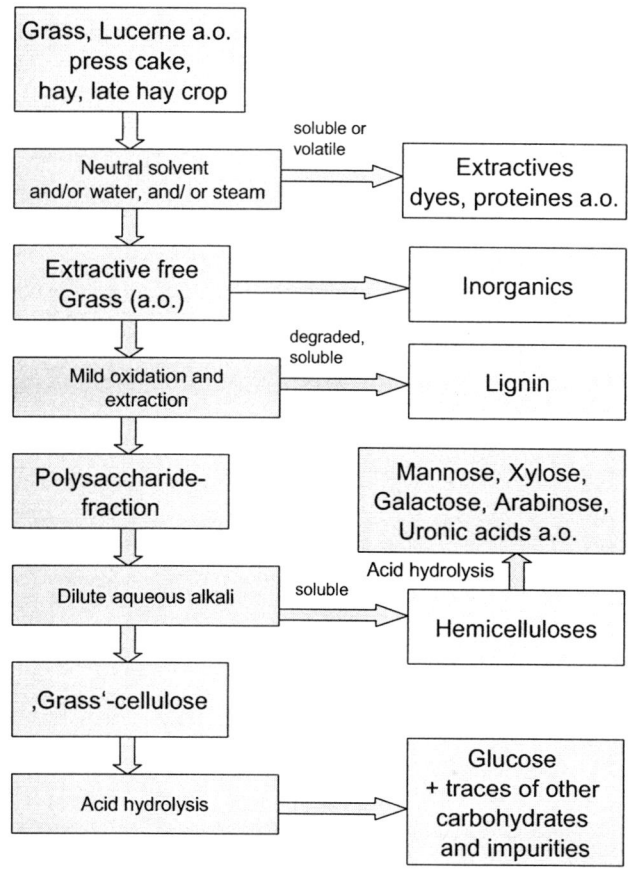

Figure 4. Classification of the major components of grass press cake, hay and late hay crop (in analogy to Janes R L, 1969).

The 'green plant' polyoses (hemicelluloses, pseudocelluloses, polysaccharides) are nutrient-physiologically valuable (Authors group,1978). Furthermore, they can be used (similar to plant rubber) as protecting colloids, emulsifiers in cosmetics, thickeners in food industry (Aspinall, 1983a), adhesives, additives in pulp and paper industry, stabilizers for environmental friendly inks and dyes (Aspinall,1983b), or as thickeners for crude oil drillings

Davidson,1980). Lignin is one of the components of lignocellulose. Isolated lignin can be used as dispersant in food industry, as stabilizer for foams and bitumen or as environmental friendly adhesive (ACS,1989; Perl,1967; Crawford,1981).

3.1. Cellulose - High Valuable Products

Highly pure plant celluloses isolated according to the process shown in Figure 4 can be used as support for immobilization and chromatographic purification of proteins (immune bodies, antigens, enzymes, lectines) (Boeden, 1991; Baeseler, 1992) or as cellulose derivatives for polyelectrolyte complex microcapsules of biological materials (proteins, enzymes, cells) (Dautzenberg et al., 1996, Dautzenberg et al., 1999, Ulrich et al. 2001). Recent development of cellulose sulfate (polyanionic component) and polycarnitine (polycationic component) based polyelectrolyte complex microcapsules are for special interests for application in medicine, pharmacy and cosmetics (Kamm et al., 2001, Kamm et al., 2005).

3.2. Cellulose - Low Valuable Products

- One of the short-term applications could be the use of the press-cake-fraction for the production of rough paper, wallpaper, cardboard and moulded articles for protective materials, (egg packings etc.) (Holm-Christensen, 1989; Fechner and Hertwig 1994). Furthermore, additives for composite materials (fiber-enhanced polymers) or conventional cellulose industry. Paper from press cake of lucerne has been manufactured in Denmark (Carlsson, 1993).

Two interesting developments should be mentioned: First, the studies to produce paper out of Reed canary-grass (Phalaris arundinacea) and cock's-foot (Dactylis gromerata) according to a conventional CTMP-process (Chemo-Thermo-Mechanical Pulp). In this respect, the Reed canary-grass (Phalaris arundinacea) from eastern German lowland moor's shows particularly good paper properties (Figure 5).

Second, the prenacellR-process regarding the production of raw paper or tech paper from late harvested grass from nature preserve area and waste lay from filament garn industry. Using the so-called yellow lye (17% waste NaOH) under water-steam pressure the fiber is breaked to the grass pulp and following is pressed to boards by filter pressing. It could be shown, that grass card boards have the same or even a better quality (for the paper re-working industry) than analogue waste papers and in addition, they are less expensive (price for waste paper: 100-140 Euro/tons) (Hille Ch, 1999). The prenacell-pulp can also be added to polymeric foam-forming compounds, which can be foamed and cured via micro-wave technology to produce insulation sheets for construction industry (based on grass).

- A further product group produced from press cake are oxygen-containing chemicals, which are bio-technologically synthesized via fermentation or from pentose/hexose. For this purpose, PC-carbohydrates (e.g. cellulose and hemicellulose) will be degraded via fermentation or chemically to monosaccharides (saccharification). This can be done stepwise or even in one step. In this way, basic chemicals such as lactic acid, ethanol, glycerine, acrylic acid and 1,3-propanediol are bio-technologically accessible in which technological access varied strongly (Danner et al. 1997).

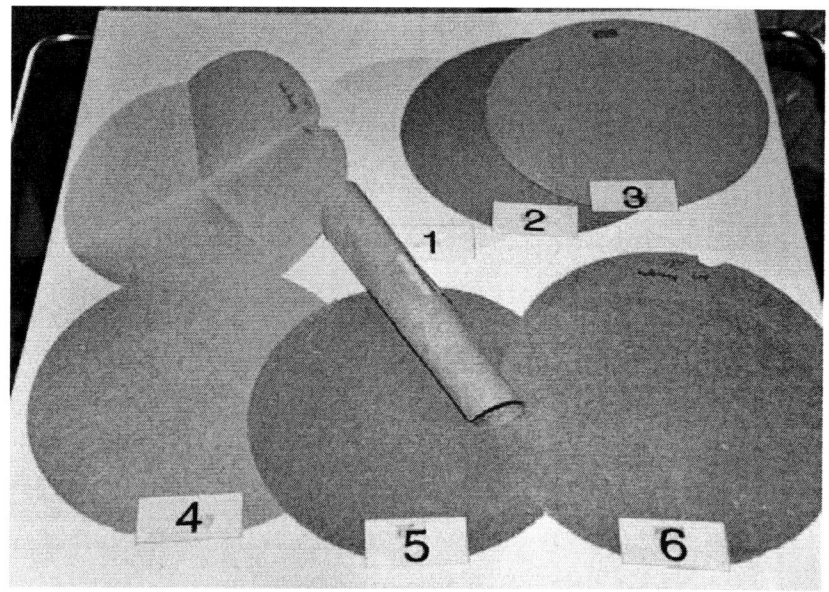

1• Wood,
2• Waste (old) paper,
3• Waste (old) paper,
4•*Phalaris arundinacea*
    - Reed canary-grass,
5• Wild mixed grass,
6• *Dactylis gromerata*
    - Cock's-foot

Figure 5. Paper made from grass, CTMP-process.

3.3. Levulinic Acid and Sequence Products

A major role as high-valuable chemical compound obtained from biomass is levulinic acid, which can be obtained from hexoses and pentoses. During the conversion, 5-hydroxymetylfurfural (HMF) is formed via acid catalyzed dehydration, which can be split via dehydration into levulinic acid and methanoic acid (formic acid). Even with raw materials that strongly vary in their quality from batch to batch, the yields that can be obtained are very high. The formic acid can be removed via distillation for further use. Levulinic acid is a versatile chemical intermediate (Dahlmann, 1968, Kuster, 1990, Olson, 2001) (Figure 6).

To decrease the waste problems, the State New York (U.S.A.) has built two levulinic acid pilot plants (1 t per day). In these pilot plants, different possibilities shell be tested for the exploitation of the whole variety of carbohydrate-rich and humid waste materials (waste paper, sewage sludge) for levulinic acid production. Thermo-chemical processes tolerate fluctuations in feedstock compositions. In the future, decentred facilities are planned with volumes of 50 to 1000 t/day and more. The high prices of levulinic acid has inhibited large-scale use. It currently has a world wide market of about one million pounds per year at a price of $4 to $6 per pound. The New York "*biofine*-process" is projected to be capable of producing levulinic acid at $0.04 to $0.32 per pound, depending on the scale of operation NYSERDA, 1998, Fitzpatrick, 1999.

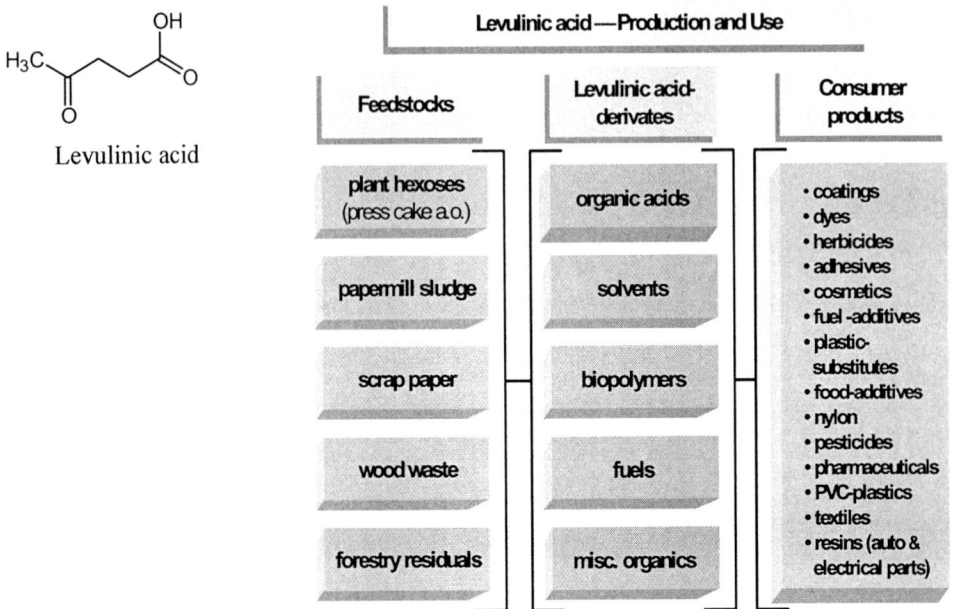

Figure 6. Levulinic acid—production and use.

- The PC has also been used as media for growing mushrooms, mulch/green crop enhancer and fertiliser .

Actually, the PC can be used as a renewable sources of carbohydrates for multipurpose uses in the same way as crude mineral oil.

4. PRODUCTS FROM THE JUICE FRACTION (GJ-GREEN JUICE, BJ-BROWN JUICE)

In the (esp. fresh pressed) GJ we can find proteins, lipids, glycoproteins, lectins, sugars, free amino acids, dyes (carotenes), hormones, enzymes, minerals and others, so especially crude drugs. The GJ can be fractionated by heat, organic and inorganic acids, acid anaerobic fermentation, centrifugation and gel filtration into a leaf nutrient concentrate (LNC) and a brown juice (BJ). The LNC consists of a mixture of chloroplastic and other organic membrane plus denaturized earlier soluble plant cell proteins. The composition of a LNC is:

true protein (60-70%); lipid: [esp. palmitic acid, linoleic acid, lionolenic acid] (20-30%), starch (5-10%), ash (1-10%), carotenoide/polyene dyes: ß-carotene (1-2 g/kg) and xanthophyll (Pirie, 1975, Schwenke, 1985). The LNC is mainly used for non-ruminant feed to enhance the colour (trough red ß-carotene) of chicken skin or egg yolk. Its also produces tender meat in chickens, ducks and pigs. Pigs fed LNC give pork with increased contents of healthy oleic and linoleic fatty acids in the fat (Carlsson, 1997).

- **Lectins** are special proteins, which are able to specifically recognise and complex polysaccharides even in lipid- or protein-bonded form. The old name Phytohäma-glutinine / phytaglutinine shows, that the first lecitins have been found in plants (e.g. wheat germs, potato's). However, not very much is known about the detailed mechanism of lectins in plants. However, there are already several applications in human and veterinary medicine (e.g. cancer diagnostics) (Berenzin et al.,1995; Pusztai and Bardocz, 1995) and plant protection (Payene et al. 1993); feed and food control (Schröder et al.,1993). Plant lecitines are available via preparative isolation and subsequent purification by means preparative chromatography.
- **Sugars** (monosaccharides, disaccharides and derivatives): GJ and BJ contain valuable special sugars and have highly valuable and sometimes expensive applications (Kirk, Othmer, 1994). Before also these sugars are fermented, they are studied regarding to their potential characterization, and isolation (Starke et al., 2000).
- **Dyes:** The GLNC enriched in ß-carotene may have anti-cancer effects. Apart from the use of beta-carotene as pro-vitamin A, ß-carotene and other carotenoid (xanthophyll) are used in cosmetic drugs and as food-, textiles- as well as toys colouring substance (see also chlorophyll (Schertz, 1938; Shearon and Gee, 1949; Judah et al., 1954).
- **Fatty acids:** The GLNC is also rich in oleic and linoleic fatty acids, especially palmitic acid, linoleic acid, ,linolenic acid. The lipids have with a good health values. The lipids can be extracted with hydrocarbons. They are interesting for the cosmetic industry (Schwenke, 1985).
- **Crude drugs/ingredients:** As the BJ contain specific secondary plant substances, such as, saponins and nicotine, these can be separated from the juice for pharmacological or pesticide purposes (for the isolation see Hagers handbook of pharmaceutical practice, 1972-78).
- **Fertilizer**: GJ or BJ used as a bio-fertiliser (soil bioactivators) to return to the soil the macro and micro mineral nutrients, which were removed by harvesting the green crop (Carlsson, 1997).

4.1. Proteins

The study of green leaf proteins (LP) goes back more than two hundred years to 1773 when Hilare Rouelle published the first known report on the subject (Rouelle, 1773). The pioneering work of Pirie (1987) since World War II focused on bulk extraction of leaf protein and possibilities for its incorporation into human diets. In the 1980's leaf protein was the main subject of three international conferences (1982 in India, 1985 in Japan and 1989 in Italy). The most intensive research on leaf proteins has been conducted with alfalpha and tobacco. A comprehensive and critical review of the plant sources, chemistry and nutrition of leaf protein concentrates as well as their preparation is given in the book "Leaf protein concentrates" (Telek and Graham, 1983). Recently research has been focused on Rubisco, the main soluble protein of the leaf (Barbeau and Kinsella, 1988; Douillard and de Mathan, 1994).

The present paper is not intended to be a definitive summary of LP research. With protein structure as the starting point, the interrelationship between extraction method, functional properties and recent and future industrial applications in the food and non-food fields will be addressed. Furthermore, experiments with other plant proteins demonstrating the potential of leaf proteins in non-food applications will be discussed.

4.1.1. Composition, Fractions and Structure of Leaf Protein From Green Biomass

On a wet weight basis the protein content of green leaves varies considerably between 1.2% and 8.2% (lucerne ~ 4%) depending upon plant species, age and growing conditions. The proteins present in leaves possess diverse structure and functions in the living plant including lipoproteins (membranes), structural proteins, photoactive pigment-bound proteins (chloroplasts), and enzymes. Leaf proteins can be divided into two classes: (i) insoluble chloroplastic leaf proteins consisting primarily of lipoproteins and pigment-bound proteins, and (ii) the water-soluble cytoplasmic fraction.

The content of soluble proteins is relatively high in the leaves of C_3 plants (33 to 48%, e.g. wheat, lucerne, tobacco) compared to C_4 grasses (26 to 30%, e.g. maize or sorghum) (Carlsson, 1995). The main soluble protein in leaves is the enzyme ribulose-1,5-bisphosphate carboxylase/oxygenase (Rubisco EC 4.1.1.39) also known as fraction-I protein. In lucerne leaves, it accounts for 30 to 70% of total nitrogen, depending on the physiological stage or genotype (Douillard and de Mathan, 1994). Rubisco has, irrespective of the plant source, a molecular weight (MW) of between 500,000 and 600,000 daltons and is composed of eight large and eight small subunits with MWs of about 55,000 and 12,500, respectively. The sedimentation coefficient is close to 18.5S. Rubisco possesses a compact, tightly folded three-dimensional structure typical of globular proteins. Due to its amino acid composition Rubisco is mildly acidic and carries a negative charge at neutral pH (isoelectric point pH 4.4-4.7) (Bahr et al., 1977). It also has a relatively high average hydrophobicity value of 1275 cal/g residue calculated according to Bigelow (1967). Native Rubisco from alfalfa contains 90 sulfhydryl groups, of which eight are "free" (one per protomer), 36 one exposed after denaturation by SDS and 46 are involved in the formation of disulfide bonds within the Rubisco subunits (Hood et al., 1981). The denaturation temperature of lucerne Rubisco varied between 70°C (pH 7.5) and 61°C (pH 10.3) (Burova et al., 1989). For more details about Rubisco, the reader is referred to the reviews of Barbeau and Kinsella (1988) and Douillard and de Mathan (1994).

Fraction-II protein consists of a mixture of proteins originating from the chloroplasts and cytoplasm with molecular weights from 10,000 to 300,000 daltons and has a sedimentation constant of from 4S to 10S (Jones and Mangan, 1976).

On the basis of the amino acid composition, Rubisco as well as the green and white fraction of leaf protein are considered hydrophobic (Table 3).

Table 3. Comparison of the amino acid composition of the different protein fractions of Lucerne (Hatch and Bruce, 1968, Douillard, 1985)

Protein	Amino acids (in parts per thousand)				
	Hydrophilic (H)	Charged	Apolar (A)	Small	H/A
Rubisco	414	289	285	180	1.45
White protein	421	303	272	183	1.55
Green protein	432	286	268	180	1.61
Soluble protein	491	333	239	143	2.10
Oligomeric soluble protein	451	310	275	166	1.70
Membranous protein	427	288	299	169	1.40

Hydrophilic: Asp + Glu + Ser + Thr + Arg + Lys + His;

Charged: Asp + Glu + Arg + Lys;

Apolar: Val + Ile + Leu + Phe + Met;

Small: Gly + Ala

4.1.2. Extraction Methods of Leaf Protein

A review of the various methods to prepare leaf protein concentrates (LPC) has been published by Hernandez-Garcia and Martinez-Para (1988). Depending on the methods used, it is possible to obtain (i) an unfractionated leaf protein concentrate (whole LPC), (ii) a decolorized whole LPC, (iii) the green chloroplastic fraction or (iv) the white cytoplasmic fraction.

The unfractionated leaf protein concentrate (whole LPC) is defined as the complex precipitated from the green juice without further fractionation, containing both the chloroplastic and cytoplasmic fractions. Such a complex is typically dark green and contains from 50 to 65 wt% protein, 15 to 30 wt% lipids, chlorophylls, carotenoids (xanthophylls), and approximately 5 wt% ash. These concentrates can be easily prepared by heat coagulation at 85-90°C using direct steam injection followed by centrifugation to separate the curde and drying (ProXan process, Knuckles et al., 1972). Due to the similar isoelectric point of both fractions, acid precipitation or anaerobic fermentation of the juice leads to whole LPC, as well (Ohshima et al., 1997). Hernandez et al. (1988) described the preparation of unfractionated LPC by freezing alfalpha juice at –25°C. The subsequent thawing at room temperature resulted in a freezing curde which can be separated by centrifugation and contains 50% of the dried matter and 60 wt% of the nitrogen present in the original juice. Ultrafiltration of the whole herbage juice using membranes with a 1 to 2×10^4 MW cut-off was suitable for obtaining a whole LPC with good solubility, but the process was time- and energy-consuming (Ostrowski-Meissner, 1983a).

In order to remove the green color and increase the storage stability of the LPC, extraction with polar solvents has been described (Bray et al., 1978). Although effective, very large amounts of solvents were necessary making this process less economical.

Fractionation of chloroplastic and cytoplasmic proteins can be achieved by exploiting their different physico-chemical properties. A well-known method (the ProXan II process) is based on the differential heating of green juice (Edwards et al., 1975). In the first step the green chloroplastic fraction containing 50 wt% proteins, 14 wt% lipids, carotenes and xanthophylls was extracted by steam injection at 60°C and centrifugation. After removing traces of green particles by filtration, the white proteins can be recovered from the clarified juice by heat (80°C, Edwards et al., 1975), acid precipitation (pH 4, Miller et al., 1975), ultrafiltration, or gel filtration (Knuckles et al., 1975). These procedures influenced considerably the functionality of the resulting LPC.

A two-stage ultrafiltration procedure using membranes with a 6.5×10^4 MW cut-off (which retained approximately 75-80 wt% of the total recoverable proteins) and a 6×10^3 MW cut-off led to a fractionation similar to differential heating (Ostrowski-Meissner, 1983a). The most advantageous method of protein separation appeared to be steam coagulation for chloroplastic fraction recovery followed by membrane filtration to separate the cytoplasmic fraction.

The application of high molecular polyelectrolytes (cationic or anionic flocculants) allowed for the separation of the pigmented chloroplastic fraction at room temperature (Knuckles et al., 1980a, Baraniak et al., 1989). Antonov and Tolstoguzow (1990) showed that linear polysaccharides (e.g. 0.55% pectin and methylcellulose) were most effective in precipitating chloroplast particles which can then be easily separated by centrifugation. After a short alkali treatment (pH 11.5 for 1 min) a completely water-soluble white protein concentrate was obtained by subsequent acid coagulation and washing.

Organic solvents with limited water solubility (e.g. butanol, ethylacetate) were most effective at a concentration of 3 wt% whereas highly polar solvents (e.g. acetone, isopropanol) should be used at 10% in order to remove the chloroplastic fraction (Bray and Humphries, 1978). Optimal solvent extraction was achieved at pH 6.0 and 25°C.

In Hungary the Vepex-(vegetable protein extract) process was developed which involves the preparation of a protein concentrate from plant juices, fermentation of the remaining juice to obtain a yeast for feed, and preparation of a feed meal from the plant tissue residue (Hollo and Koch, 1978). The fractionation process, which aimed to obtain the largest amount of white protein fraction possible, was carried out by separating the chloroplast fraction at lower temperature (40-50°C) in the presence of polyvinylpyrrolidone or Ca^{++}, Al^{+++}, and/or Fe^{++} flocculants, or at 55°C with the addition of surfactants, to delay the separation of the white protein fraction (Koch, 1983).

Tobacco is the only species from which crystalline Rubisco has been so far extracted in significant amounts (Pedone et al., 1995).

Recently a process leading to a protein isolate rich in Rubisco was described (Levesque and Rambourg, 2001). The green juice of alfalfa was separated into (i) a brown juice rich in Rubisco and (ii) a fraction rich in proteins, pigments, vitamins, insolubles and trace elements. Fraction (ii) was subsequently dried to from a food supplement. Rubisco in fraction (i) was acid-precipitated, concentrated, purified by microfiltration and diafiltration, and dried. Alternatively, purification of the brown juice was accomplished with a resin adsorber.

4.1.3. Optimization of the Protein Extraction Process

A comprehensive review of the factors limiting protein recovery and their overall effect on the yield is given by Ostrowski-Meissner (1983b). The factors were grouped into four categories: (i) ecology and agronomy, (ii) plant harvesting, (iii) plant processing and (iv) juices processing.

Some important factors will be mentioned. The maximum yield of leaf protein per hectare resulted from cuts of regrown plants, where each regrowth of the plant has reached the physiological stage of prebloom (Carlsson, 1995). With respect to the processing, the pressure conditions were important. To get maximum yields of LP it is favorable to apply pressure slowly so as to extract as much juice as possible with gentle pressure. Any delay in processing should be avoided. Delays can cause an increase in degradation by proteolytic enzymes and an increase in the binding of phenolic compounds to proteins (Pirie, 1994). Metabisulphite or sulfite salts are commonly used in leaf protein isolation to inhibit polyphenoloxidases, improve retention of methionine and available lysine, and increase leaf protein yields.

4.1.4. Applications of the Proteins

A prerequisite for exploiting proteins in special applications is the tailored modification of their functional properties. In the case of food, these properties have been defined as the "intrinsic physicochemical characteristics which affect the behavior of proteins in food systems during processing, storage, preparation and consumption" (Kinsella, 1979). Functional properties can be described as the ability of the protein to emulsify, to adsorb water, to form gels, to provide viscosity and elasticity, or to foam. In addition to the intrinsic attributes of the proteins (i.e. composition, amino acid sequence, conformation and structure), the functionality of proteins is affected by interactions with other components present (i.e. carbohydrates, lipids, salts, surfactants) and the process parameters (i.e. temperature, pH, ionic strength, reducing agents, storage conditions) (Kinsella, 1979). Several techno-functional properties of proteins, their required physico-chemical and molecular parameters and the consequences for different industrial applications are summarized in Table 4.

4.1.5. Nutritional and Functional Properties of Leaf Protein

To date, the described utilization of leaf protein concentrates has been limited to feed and food applications. The whole leaf concentrates or chloroplastic fraction are primarily used for feeding nonruminant animals and poultry. The high pigment content (e.g. xanthophylls) of LPC distinguishes it from the products based on soybeans. Broiler and layer feeding trials have shown that the xanthophylls in Pro-Xan are utilized more efficiently than the xanthophylls in dehydrated alfalfa (Kuzmicky et a.,1977). LPC included in the diet of laying hens increased their egg production slightly, and also increased the intensity of the yolk color of the eggs. The aim of a recent Austrian project was to compare the use of fresh green protein and silage protein as a specialty feed (milk re-placer) for calves (Koschuh et al., 2003). Whereas high quality protein products were obtained from fresh biomass, the yield of silage protein was too low due to a high degree of proteolysis.

On the basis of amino acid composition, LPC showed high nutritive values. The limiting amino acids in leaf proteins are the sulfur-containing amino acids methionine and cysteine. Depending on the production method, the digestibility of whole LPC is 80-90%, while for cytoplasmic LPC it is usually higher than 95%. The nutritive value is affected adversely by the presence of antinutritional factors (e.g. amino transferase activity, trypsin inhibitory activity, tannins, saponins) which differed dependent on the plants and production process (Hussein et al., 1999).

A review of the investigated functional properties of leaf proteins in food applications has been written by Barbeau (1990). It is obvious that the extraction method has considerable impact on the properties and suitability of the leaf protein preparations in food systems.

Table 4. Relationship between physico-chemical and techno-functional properties and the consequences for different industrial applications

Techno-functional properties	Physico-chemical properties Molecular and structural prerequisites	Suitable modifications (selection)	Industrial applications (examples)
Emulsification	Amphiphilic character, high solubility, ease of unfolding, high molecular flexibility and hydrophobicity	Partial hydrolysis (DH<10, MW>10kDa); acetylation, succinylation	Preparation of stable oil-in-water emulsions
Foaming	High solubility, surface hydrophobicity, ease of unfolding, reorientation at the interface without aggregation or coagulation; decreased molecular weight; enhancement of protein-protein-interaction (electrostatic) to decrease lipid-sensitivity	Partial hydrolysis (DH<10, MW>10kDa); physical protein degradation (mechanolysis)	Preparation of aerated solutions, products and foams
Gelation	Long, coiled polypeptides capable of loosing and unfolding by heat; deposition of charged groups; ordered reassociation via hydrophobic association, hydrogen bonding, ionic interactions and/or disulphide linkages	Acetylation, alkali and heat treatment, carbonyl-amine reaction, reaction with bivalent ions	Gel formation (food gels, emulsion gels, surface coatings, film formation, membranes stable to organic solvents)
Water adsorption and holding	Reduced solubility for physical water binding	Chemical modification by acylation and cross-linking of the acyl-modified protein with dialdehyde (*)	Water binding powders, textiles, absorber
Viscosity, thickening power	Increased hydrodynamic properties (molecular volume, size, axial ratio, shape of the molecule) caused by denaturation without gelation	Alkali treatment (> pH 9)	Thickener, stabilizer (dispersed systems), regulation of flow behavior of coatings or building materials
Fat adsorption Adhesion;	High surface hydrophobicity Unfolding (flat conformation, less structured), high polar affinity to the surface, exposure of specific groups,	Heat denaturation and drying Cross-linking to enhance cohesion and reduce water sensitivity	Binding of nonpolar liquid contaminations Adhesive, coatings
Cohesive strength Metal-binding	Unfolding, exposure of specific groups	Chemical modification by acylation and cross-linking of the acyl-modified protein with dialdehyde (*)	Industrial absorbents, waste water reclamation, heavy metal sequestration

Kinsella (1979), Feeney (1987), Muschiolik (1991), De Graaf (2000)

* Damadoran & Hwang, 1995

Good solubility plays a key role in the quality of LPC because it is a prerequisite for emulsification, foaming or gelation. The frequently used method of heat coagulation results in irreversible denaturation and loss of solubility. Using the differential heating method, both green chloroplastic and white cytoplasmic alfalfa protein concentrates showed poor solubility between pH = 2 and 10. On the other hand, acid-precipitated leaf protein was soluble above pH = 6 and below pH = 3 (Betschart and Kinsella, 1973). Removal of phenolic compounds by ultrafiltration resulted in increased solubility of alfalfa protein concentrate (Knuckles et al., 1980b). The emulsifying activities and emulsion stabilities of acid-precipitated alfalfa leaf protein concentrates were better than those of soybean concentrate (Wang and Kinsella, 1976a). Extraction of the lipids with acetone led to a slight decrease in solubility and a 50% reduction in water and fat adsorption capacities compared with controls. Both foam formation and stability, however, were markedly improved by extraction of the lipids (Wang and Kinsella, 1976b). The addition of sucrose and NaCl reduced the foaming of the leaf proteins. The edible protein concentrate from alfalfa prepared by ultrafiltration and spray-drying at 85°C showed excellent functional properties (Knuckles and Kohler, 1982). Emulsions containing 2 wt% alfalfa protein and oil were stable and their consistency was similar to mayonnaise. Firm gels were obtained by heating solutions of 3 wt% alfalfa protein to 72°C and cooling. Foam volume and stability were similar to those of egg white.

Since the 1980's much research has focused on the purification and technological evaluation of Rubisco, the main soluble protein of leaves. Due to the excellent functional and nutritional properties of this tasteless and odorless white powder, purified Rubisco is regarded as having great potential as an ingredient in animal and/or human foods. The reviews of Barbeau and Kinsella (1988) and Douillard and de Mathan (1994) demonstrated that Rubisco was able to form stable foams and emulsions at low temperature, to produce firm gels when heated and to give high fat-binding capacity when the protein has not been modified by heat treatment.

4.1.6. Effects of Enzymatic and Chemical Modification of Leaf Proteins

The modification of proteins via enzymatic, chemical or physical procedures is a common way to tailor the functionality. It is generally accepted that limited hydrolysis enhances the surface properties, whereas extensive hydrolysis has a detrimental effect on the stabilization properties. It was shown that the initial enzymatic degradation of native crystalline tobacco fraction-1 protein led to the removal of oligopeptides from the large subunits without a loss of small subunits and destruction of the quaternary structure (Sheen and Sheen, 1987). Insoluble alfalfa protein concentrate was solubilized by the proteolytic enzyme Delvolase and a peptide isolate suitable as a protein supplement in human diets was obtained by coupling the process with ultrafiltration using a 10 kDa MW cut-off membrane (Prevot-D'Alvise et al., 2003). According to Yang et al. (2004) the pepsin digest of leaf protein with a high Rubisco content could potentially be useful in the prevention and/or treatment of high blood pressure (hypertension). So far, the influence of the degree of hydrolysis on the functional properties of leaf protein samples has not been investigated.

Chemical modification can be subdivided into (i) hydrophilization (incorporation of polar groups, e.g. –COOH, -NH₂, -OH), (ii) hydrophobization (incorporation of apolar groups, e.g. alkyl or aromatic) and (iii) cross-linking (covalent linking of protein molecules). Only a few studies exist regarding the chemical modification of leaf proteins. The succinylation of 84% of the ε-amino groups of lysine resulted in improved flavor, increased solubility (>10-fold), enhanced emulsifying activity (32%) and increased foaming capacity (3-fold) of protein isolated from alfalfa leaves (Franzen and Kinsella, 1976). Acetylation of leaf protein improved the solubility and foaming capacity also but to a much lesser extent. Studies of the soluble tobacco leaf proteins (F-1-p and F-2-p) confirmed an improvement in functionality by succinylation (Sheen, 1991).

4.1.7. Future Trends and Necessary Investigation

In order to force the valorization of biomass components, it is important to develop a new market for leaf proteins. Besides the food and feed field, technical applications of proteins are a promising area. Such non-food applications cover a very heterogeneous market ranging from surfactants (emulsifiers, detergents, wetting agents), coatings (e.g. in paint, ink, paper, packaging), and adhesives to biodegradable plastics or materials (e.g. disposable dishes or cutlery). The market potential of renewable materials depends primarily on their superior performance, including functionality and price. The suitability of leaf proteins for such non-food applications has been insufficiently tested. In most cases, the presence of antinutritive components, dark color or unpleasant grassy smell is not prohibitive to the use of leaf proteins in non-food products. The excellent interfacial and gelling properties of a properly extracted cytoplasmic fraction and Rubisco make them suitable for use as surfactants or film-forming agents in personal care products or biodegradable packaging films. The extreme purity of crystallized leaf protein (e.g. Rubisco from tobacco) may support its use in the field of medicine and pharmaceutics. Due to its large charge/mass ratio, Rubisco has a high tendency to bind ionic substances (Barbeau and Kinsella, 1988). This may allow it to be used as a sequestering agent for environmental pollution.

Although for food applications it is important that the proteins are in the native (not denatured) state in order to exert the desired functionality, for technical applications this is less relevant (de Graaf, 2000). The basis for the formation of biomaterials is the reactivity of the macromolecules. The ability of water-insoluble plant proteins to produce biodegradable materials with antistatic and flame resistant properties (useful in electrical devices) will be described in order to demonstrate the potential of leaf protein in such applications. During a simultaneous treatment of pressure (1.2 MPa) and temperature (120 – 170°C) the raw materials (e.g. plant protein with 25% water or a mixture of plant proteins, native potato starch and/or dialdehyde starch with 25% water) were plasticized, fluidized and cooled to create materials with special mechanical properties (Bindrich, 2004). Different plant protein sources when used to make "bioplastics" demonstrate differences in mechanical behavior which cannot be predicted solely from the molecular status of the protein samples (Figure 7). Under certain conditions, the mechanical properties (bending elastic modulus) of composite materials made with wheat were superior to plexiglass (bending elastic modulus 1000 MPa) (Figure 8).

At temperatures above 120°C it is so far impossible to predict changes within the protein molecules, i.e. breaking or forming of bonds, and their effects on the mechanical properties of biodegradable materials. Only experimental tests can provide the desired information. Hence, leaf protein concentrates (single or combined with other plant proteins or starch) have potential in the production of biodegradable materials. To date, nearly all information on protein denaturation has been obtained from research on solutions with water content > 90 wt%, and little research has been done in systems with water content lower than 30 wt%. The denaturation temperature of proteins strongly depends on the water content, and can increases to 120-200°C at water contents < 20 wt% (de Graaf, 2000). By using green leaf proteins in the production of packaging materials, the existing pigments (e.g. chlorophyll) could be used to protect light-sensitive substances from oxidation.

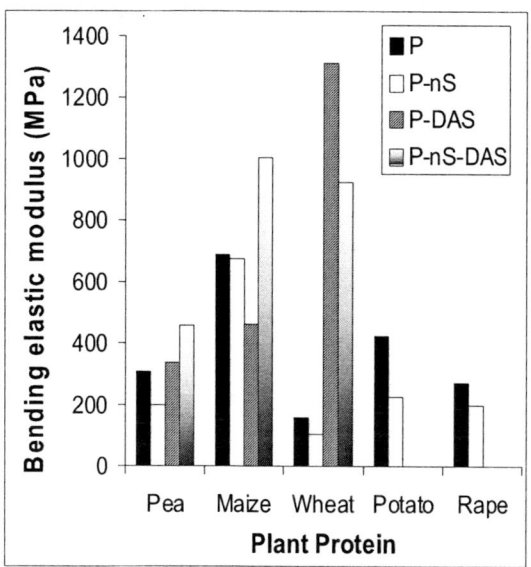

Figure 7. Mechanical properties of biomaterials made from protein (P) or a mixture of proteins, native starch (nS) and/or dialdehyde starch (DAS).

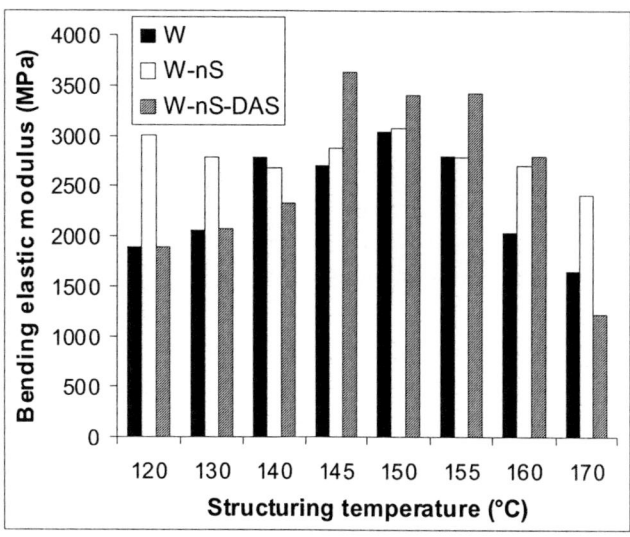

Figure 8. Effect of structuring temperature on the mechanical properties of biomaterials made from wheat protein II (W) or a mixture of proteins, native starch (nS) and dialdehyde starch (DAS).

The tailoring of protein functionality by physical, chemical and enzymatic methods is a promising method for developing new non-food applications and should be utilized more extensively for leaf proteins. Due to numerous potential glutaminyl and lysyl sites, Rubisco (mostly its large sub-unit) is a good substrate for transglutaminase-directed modifications

(Margosiak et al., 1990). This offers new opportunities for solvent stable membranes, biodegradable films and "bioplastics" with different mechanical firmness.

4 2. Green Juice and Brown Juice as an Fermentation Media and the Products

The GJ and BJ are an excellent fermentation media, as such, and mixed with cereal grain flours, molasses or other hexosen-riches substrates (Thomsen et al., 2004). As fermentation medium, BJ has been used for production of ethanol, lactic acid and other organic acids, plus feed proteins, lysine, soil bioactivators. Especially the fermentation products ethanol, lactic acid and other organic acids, amino acids such as lysine are high-values chemical Biorefinery products. Together with the PC levulinic acid these chemicals parent products and substances can be able to establish a biomass industry and a modern green chemistry.

4.2.1. Lactic Acid - Sequence Product Poly(Lactic Acid)

Lactic acid has the potential to become a commodity chemical. High growth rates are expected: an annual volume of 1.36 to 1.8 million tons for lactic acid sequence products alone for the U.S.- market (Gruber, 2001).Chemical products of sequence are propylene glycol, propylene oxide, and epoxides. Propylene oxide is a starting material for the production of polyester, polycarbonates, polyurethanes. Further products are acrylic acid as monomer for polyacrylic acid and resins as well as alkyllactates for application as „green solvents". Furthermore enantiomeric forms of lactic acid are applied in drugs, pharmaceuticals and agrochemicals. Classical areas of application are such as so-called *culinary delight* lactic acid, e.g. in food industry and technical processes such as tanneries, textile industry, chemical industry (Kamm et al, 2001).

Enormously growing amounts are expected for polymeric materials of lactic acid, poly(lactic) acid. Polylactide is a versatile thermoplastic, which can be processed in manifold ways: e.g. spinning fibres, melting spinning fibres, extrusion foils, injection moulding, thermoforming sheets, extrusion coating for paper and board and many other applications. It is fully biodegradable and compostable and does not disturb the normal process of biodegradation in compost.

Especially for the market segments of food packaging and food service, e.g. disposable - articles) and performance-products for agriculture as well as fibres for textiles PLA is an very interesting material.

In 1993, a market volume of 140.000 – 900.000 tons per year for biodegradable polymers on basis of lactic acid was estimated for the U.S.A. (Cargill,1993). Since then efforts were increased to build major industrial capacities. The company Cargill Dow LLC has built a commercial production facility for polylactide (PLA) in Blair Nebraska, U.S.A.. The Blair facility started its operations in late 2002 and has a maximum capacity of 140,000 metric tons of PLA per year (Hovey, 2002). The establishment of further capacities of the company shall follow within the next 10 years up to a capacity of 450,000 metric tons in Asia and Europe. So it is expected that the price will decrease within the next years to the level of petrochemical based thermoplastics (Cargill Dow, 2001).

The *Polylactid technology* starts with the classical variant of lactic acid fermentation (primary production of inorganic salts of lactic acid) via further steps of production of lactic acid and oligomeric lactic acid, subsequently synthesis of cyclic diesters of lactic acid, from which the polylactide is accessible by ring opening reaction. In this technology principle the two processing steps (biotechnological production of lactic acid and the subsequent chemical steps) are clearly separated (Kamm and Kamm, 1998).

A new technology principle is patented and briefly drafted in Figure 4 (Kamm et al. 1996, Kamm et al. 1997). Aminium lactates act as alternative coupling reagents to connect biotechnological and chemical conversion (synthesis of dilactide). Requirements on amium lactates are low melting points, good crystallinity and good thermal or hydrolytic dissociation ability. Requirements on amines are good water solubility, a pH-value within basic area, sufficient thermal, acid and catalyst stability for the recycling according to the cyclization process, ecological harmlessness and economical availibility. Piperazine dilactate is for example a well investigated model substance (Kamm et al, 2001), (Figure 9).

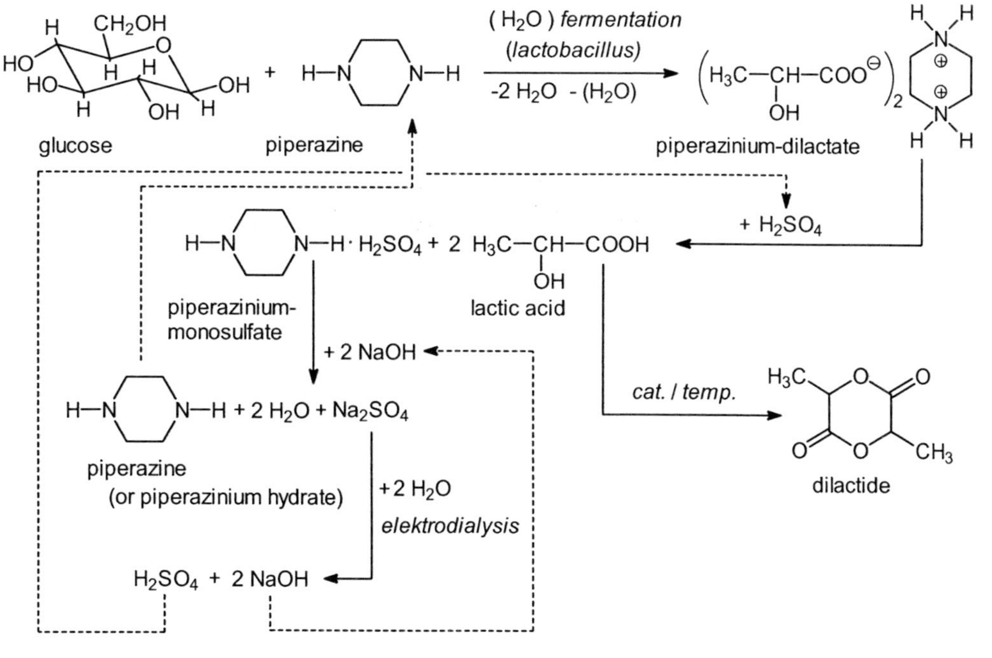

Figure 9. Principle of procedure of extremely pure lactic acid and of dilactide based on aminium lactates.

REFERENCES

ACS (1989) (American chemical society). *ACS Symp.* Ser. No. 397.

Antonov YuA, Tolstoguzov VB (1990). Food protein from green plant leaves. *Nahrung* 34:125-134.

ASPINALL (1983), (ed.),*The polysaccharides*. Vol. 2 [Academic Press, New York a) pp. 98-193, b) pp. 411-490.

Authors group (1978), *Deutsches Ärzteblatt*, 75, 2735-2741.

Baeseler M, Boeden H F, Koelsch R, Lasch J (1992), Cellulose beads: A weak leaking affinity support. *J. Chromatogr.* 589, 93-100.

Bahr JT, Bourque DP, Smith HJ (1977). Solubility properties of fraction I proteins of maize, cotton, spinach, and tobacco. *J Agr Food Chem* 25:783-789.

Baraniak B, Baraniak A, Bubicz M (1989). Protein concentrate from alfalfa and cocksfoot by polyelectrolyte precipitation. *Nahrung* 33:491-495.

Barbeau WE (1990). Functional properties of leaf proteins: Criteria required in food application. *Ital J Food Sci* 2:213-225.

Barbeau WE, Kinsella JE (1988). Ribulose bisphosphate carboxylase/oxygenase (Rubisco) from green leaves – potential as a food protein. *Food Rev Int*41:93-127.

Berenzin et. al. (1995); The Sorbents for Lectin Fraction into Isolectins. *Appl. Biochem. Microbiol.*, 3 254-258.

Betschart A, Kinsella JE (1973). Extractability and solubility of leaf protein. *J Agr Food Chem* 21:60-65.

Bigelow CC (1967). On the average hydrophobicity of proteins and the relation between it and protein structure. *J Theor Biol* 16:187-211.

Bindrich U (2004). Werkstoffe aus Protein-Kohlenhydrat-Verbunden. *Proceedings 10th International Conference for Renewable Resources and Plant Biotechnology*, Magdeburg, Germany.

Boeden H F, Pommering K, Becker M, Rupprich C, Holtzhauer M, Loth F, Müller R, Bertram, D (1991), Bead cellulose derivatives as support for immobilization and chromatographic purification of proteins. *J. Chromatogr.*, 552 389-414

Bray WJ, Humphries C (1978). Solvent fractionation of leaf juice to prepare green and white protein products. J Sci Food Agr 29:839-846.

Bray WJ, Humphries C, Ineritei MS (1978). The use of solvents to decolorize leaf protein concentrate. *J Sci Food Agr* 29:165-171.

Burova TV, Soshinskii AA, Danilenko AN, Antonov YuA, Grinberg Vya, Tolstoguzov VB (1989). Conformation stability of ribulose diphosphate carboxylase of alfalfa green leaves according to the data of differential scanning microcalorimetry. *Biofizika* 34:545-549.

Cargill-Dow (2001) Cargill-Dow to Up PLA Capacity to 450,000 t/y in 10 Yrs. Japan *Chemical Week*, 11.10.2001.

Carlsson R (1989) Green Biomass of Native Plants and new Cultivated Crops for Multiple Use: Food, Fodder, Fuel, Fibre for Industry, Photochemical Products and Medicine. In: *New Crops for Food and Industry* [Wickens et al., Chapmann and Hall, London].

Carlsson R (1993) Pressed crop for possible production of paper crop. *Proc. 4th int. Conf. Leaf Protein Res.*, New Zealand-Australia, 69-74.

Carlsson R (1995). Sustainable primary production - green crop fractionation: Effects of species, growth conditions, and physiological development on fractionation products. In *Handbook of plant and crop physiology* (Ed. Pessarakli M), Marcel Dekker Inc., N.Y., USA, 941-963.

Carlsson R (1997) Food and Non-food uses of immature cereals. In: *Cereals —Novel Uses and processes* [G.M. Campbell, C. Webb & S.L. McKee (eds.), Plenum Publ.Corp., New York] 159-67.

Carlsson, R (1985) An ecolocically better adapted agriculture. Wet fractionation of biomass as green crops. Macro alga and tuber crops. [*Proc. 2nd int. Conf. Leaf Protein Res.*, Naoya, Japan].

Carlsson, R (1998). Status quo of the utilization of green biomass In: *The Green Biorefinery, concept of technology.* (First international symposium on green biorefinery) Neuruppin, Society of Ecological Technology and System Analysis, Berlin, ISBN 3-929672-06-5] pp. 39-45.

Coombs J, Hall K (1997) The potential of cereals as industrial raw materials: Legal technical, commercial considerations. In: *Cereals - Novel Uses and processes* [G.M. Campbell, C. Webb & S.L. McKee (eds.), Plenum Publ.Corp., New York,] 1-12.

Crawford (1981) (ed.), *Lignin Biodegradation and Transformation.* [John Wiley & Sons, New York].

Dahlmann J (1968); Notiz über die Darstellung von Lävulinsäure. *Chem Ber*, 101, 4251-53.

Damodaran S, Hwang D-C (1995). *Carboxyl-modified superabsorbent protein hydrogel.* US 5,847,089.

Danner H, Boller V, Neureiter M, Sandhofer M Braun R (1997) Neue biotechnologische Verfahren zur Produktion marktrelevanter C$_2$-, C$_3$- und C$_4$-Verbindungen aus Biomasse. In: *Chemie nachwachsender Rohstoffe. Proc. of the conf,.* Wien, [BMUJF (ed.), Wien, Austria, ISBN: 3-901 305-71-8].

Dautzenberg et al. (1999) Development of Cellulose Sulfate-based Polyelektrolyte Complex Microcapsules for medical applications, In: *Bioartifical Organs II, Technology, Medicine, and Materials, Annals of the New York Academy of Sciences*, Vol. 875, pp. 46-63.

Dautzenberg H, Arnold G, Tiersch B, Lukanoff B, Eckert U (1996) Polyelektrolyte complex formation at the interface of solutions, *Progr Colloid Polym Sci* 101, 149-156.

Davidson (1980), (ed.); *Handbook of Water soluble Gums and Resins.* [McGraw-Hill, New York].

De Graaf LA (2000). Denaturation of proteins from a non-food perspective. *J Biotechnol* 79:299-306.

Douillard R, de Mathan O (1994). Leaf protein for food use: potential of Rubisco. In *New and developing sources of food proteins* (Ed. Hudson BJF), Publisher Chapman & Hall, London, UK, 307-342.

Douillard, R (1985). Biochemical and physicochemical properties of leaf proteins. In *Proteines Veg,* Publisher Tech. Doc. Lavoisier, Paris, France, 211-244.

Ebert G. (1993) *Biopolymere, Struktur und Eigenschaften.* B. G. Teubner Stuttgart, (ISBN: 3-519-03516-2).

Edwards RH, Miller RE, de Fremery D, Knuckles BE, Bickoff EM, Kohler GO (1975). Pilot plant production of an edible white fraction leaf protein concentrate from alfalfa. *J Agr Food Chem* 23:620-626.

Estimates from Battelle, SRI, Cargill (1993) *announcement.*

Fechner M (1998) Grassland economy in Brandenburg, In: *The Green Biorefinery, concept of technology.* (First international symposium on green biorefinery) Neuruppin, Society of Ecological Technology and System Analysis, Berlin, ISBN 3-929672-06-5] pp.47-52.

Fechner M, Hertwig F (1994); Paper made of grass very much in the future. *Neue Landwirtschaft*, 11 29-31 (germ.)

Fechner M, Hertwig, F (2003) *State institute of Agriculture Brandenburg,* Paulinenaue, personal information.

Feeney RE (1987). Chemical modification of proteins: comments and perspectives. *Int J Pept Res* 29:145-161.

Fitzpatrick S (1999) (biofine inc.); Process for Conversion of Cellulosic Biomass to Chemicals. In: *Green Chemistry and Engineering. Proc. 3rd Ann. Conf.* Wahington, D.C. USA, june 1999 [ACS (American Chemical Society), Washington D.C.].

Franzen KL, Kinsella JE (1976). Functional properties of succinylated and acetylated leaf protein. *J Agr Food Chem* 24:914-919.

Gruber P(2001), Nature Works: New Products, Markets, and Sustainability, *Presentation* 19.12.2001, Industrial Investment Council, Berlin.

Hagers Handbuch der pharmazeutischen Chemie (1972-78) (Hagers handbook of phamaceutical practice) *Vol. 1-5, chemicals and drugs,* [W. Kern et al. (eds), Springer Verlag, Berlin, Heidelberg, New York, Vol. 1, pp.732-36

Hatch FT, Bruce AL (1968). Amino acid composition of soluble and membranous lipoproteins. *Nature* 218:1166-1168.

Hector A, et al. (1999) Plant Diversity and Productivity Experiments in European Grasslands, *Science,* 286, 11231127.

Hernandez A, Martinez C, Gonzalez G (1988). Effects of freezing and pH of alfalfa leaf juice upon the recovery of chloroplastic protein concentrates. *J Agr Food Chem* 36:139-143.

Hernandez-Garcia A, Martinez-Para MC (1988). Leaf protein concentrates. A review. *Anales de Bromatologia* 40:311-325.

Hertwig F, Scholz, V (1998), Thermal conversion from grass. In: *The Green Biorefinery, concept of technology.* (First international symposium on green biorefinery) Neuruppin, Society of Ecological Technology and System Analysis, Berlin, ISBN 3-929672-06-5] pp. 93-100.

Hille Ch (1999) *Prenacell-process,* UVER – Umweltverfahrenstechnik GmbH, Premnitz, Germany.

Hollo J, Koch L (1978). VEPEX, das ungarische Blattprotein-Verfahren. *Ernaehrungsumschau* 25:210-216.

Holm-Christensen (1989); The dehydration plant as producer for the cellulose industry. *Proc. Dry-Crops 89, 4th Int. Green Crop Drying Congr.,* Cambridge [Agra Europe Ltd., London, UK] pp. 91-94

Hood LL, Cheng SG, Koch U, Brunner JR (1981). Alfalfa proteins: Isolation and partial characterization of the major component - fraction I protein. *J Food Sci* 46:1843-1850.

Hovey A (2002), Cargill delivers $300M message with new plant. *Lincoln Journal Star* Lincoln, Nebrasca, 03.04.2002.

Hussein L, El-Fouly MM, El-Baz FK, Ghanem SA (1999). Nutritional quality and presence of antinutritional factors in leaf protein concentrates (LPC). *Int J Food Sci Nutr* 50:333-343.

Janes R L (1969); In: *The pulping of wood.* 2nd edn, vol 1 [R.G. Mac Donald (ed), McGraw-Hill, New York,] p. 34.

Jones WT, Mangan JL (1976). Large-scale isolation of fraction 1 leaf protein (18S) from lucerne (Medicago sativa L). *J Agr Sci* 86:495-501.

Judah M A, Burdick E M, Carroll, R G, (1954) Chlorophyll by solvent extraction. *Industrial and Engineering Chemistry,* 46, 2262-2271.

Kamm B (2004) *Neue Ansätze in der Organischen Synthesechemie – Verknüpfung von biologischer und chemischer Stoffwandlung am Beispiel der Bioraffinerie-Grundprodulte* Milchsäure und Carnitin, Habilitationschrift, Universität Potsdam.

Kamm B et al (2000) Green biorefinery Brandenburg, article to development of products and of technologies and assessment. *Brandenburg Umweltber* 8, 260-269.

Kamm B, Kamm M (1998)Milchsäuren als grüne Biochemikalien. In: *The Green Biorefinery, concept of technology*. (First international symposium on green biorefinery) Neuruppin, Society of Ecological Technology and System Analysis, Berlin, ISBN 3-929672-06-5] pp.89-92

Kamm B, Kamm M (2004) Mini Review, Principles of biorefineries, *Appl Microbiol Biotechnol* 64, 137-145.

Kamm B, Kamm M, Fischbach M, Allwohn J, Birkel S, Krause Th (2001) *Polyesters and oligoesters of cationic hydroxy acids, method for production them and use thereof,* PCT/WO 01/94441 A1, 11.05.2001/13.12.2001.

Kamm B, Kamm M, Kiener A, Meyer H P (2005) MINI-REVIEW, Polycarnitine – a new biomaterial, *Appl Microbiol Biotechnol*, in press.

Kamm B, Kamm M, Richter K (2001) Biobased industrial products. Heterocyclic aminium lactates. New fermentation products and feedstocks for multiple use, *Agro-Food-Industry Hi- Tech*, 12 (3), 15-19.

Kamm B, Kamm M, Richter K, et al. (1996) *Verfahren zur Herstellung von 1,4-Dioxan-2,5-dionen, neue Piperazoniumsalze von α-Hydroxycarbonsäuren und Fermentationsverfahren zur Herstellung von organischen Aminiumlactaten. Offenlegungsschrift* DE 196 28 519 A1, Cl. C07 D 319/12 (04.07.1996/ 14.08.1997) 18 S.

Kamm B, Kamm M, Richter K, et al. (1997) *Verfahren zur Herstellung von organischen Aminiumlactaten und deren Verwendung zur Herstellung von Dilactid.* Europäische Patentanmeldung EP 0 789 080 A2, Cl. C12P 7/56, (07.02.1997/ 13.08.1997).

Kamm B, Kamm M, Richter K, Reimann W, Siebert A (2000) Formation of Aminium Lactates in Lactic Acid Fermentation. Fermentative production of 1,4-Piperazinium-(L,L)-dilactate and its use as starting material for the synthesis of dilactide (Part 2), *Acta Biotechn.*, 20, (3-4), 289-304.

Kinsella JE (1979). Functional properties of soy proteins. *J Am Oil Chem Soc* 56:242-258.

Kirk R E, Othmer D F(1994) (ed.).; *Encyclopedia of chemical technology* [John Wiley & Sons, New York].

Knuckles BE, Bickoff EM, Kohler GO (1972). PRO-XAN process: Methods for increasing protein recovery from alfalfa. *J Agr Food Chem* 20:1055-1057.

Knuckles BE, de Fremery D, Bickoff EM, Kohler GO (1975). Soluble protein from alfalfa juice by membrane filtration. *J Agr Food Chem* 23:209-212.

Knuckles BE, Edwards RH, Kohler GO, Whitney LF (1980a). Flocculants in the separation of green and soluble white protein fractions from alfalfa. *J Agr Food Chem* 28:32-36.

Knuckles BE, Edwards RH, Miller RE, Kohler GO (1980b). Pilot scale ultrafiltration of clarified alfalfa juice. *J Food Sci* 45:730-734.

Knuckles BE, Kohler GO (1982). Functional properties of edible protein concentrate from alfalfa. *J Agr Food Chem* 30:748-752.

Koch L (1983). The Vepex process. In *Leaf protein concentrates* (Eds. Telek L, Graham HD) AVI Publishing: Westport, CT, USA, 601-631.

Koschuh W, Kromus S, Krotscheck C (2003). Grüne Bioraffinerie – Gewinnung von Proteinen aus Grassäften. Berichte aus Energie- und Umweltforschung 19/2003, BMVIT, Wien. http://www.nachhaltigwirtschaften.at/nw_pdf/0319_proteine.pdf

Kromus et al. (2004) The Green Biorefinery Austria – Development of an Integrated System for Green Biomass Utilization, Chem Biochem Eng Q 18 (1) 13-19

Kuster B F M (1990); 5-Hydroxymethylfurfural (HMF). A Review Focussing on ist Manufacture. starch/Stärke, 42, 314-321

Kuzmicky DD, Livingston AL, Knowles RE, Kohler GO, Guenthner E, Olson OE, Carlson CW (1977). Xanthophyll availability of alfalfa leaf protein concentrate (Pro-Xan) for broilers and laying hens. Poultry Sci 56:1504-1509.

Leblanc R M, et al.; (1984) Photobiochem. Photobiophys. 7, 41.

Levesque D, Rambourg J-C (2001). Procede pour traiter le jus vert issu du pressage d'une matiere foliaire riche en proteins telle que la luzerne. FR 2819685.

Margosiak SA, Dharma A, Bruce-Carver, MR, Gonzales AP, Louie D, Kuehn GD (1990). Identification of the large subunit of ribulose 1,5-bisphosphate carboxylase/oxygenase as a substrate for transglutaminase in Medicago sativa L. (alfalfa). Plant Physiol 92:88-96.

Meissner D (1997) Farbstoffzellen-Solarenergieumwandlung mit organischen Farbstoffen. Photon, 2, 24-27.

Miller RE, de Fremery D, Bickoff EM, Kohler GO (1975). Soluble protein concentrate from alfalfa by low-temperature acid precipitation. J Agr Food Chem 23:1177-1179.

Moser A (ed.) (1997], Eco-tech The technical Development, Verlag TU Graz, Austria

Muschiolik G (1991). Proteinprodukte als funktionelle Lebensmittelkomponenten (I). Lebensmitteltechnik 11/91: 630-636.

National Research Council, USA (2000) Biobased Industrial Products, Priorities for Research and Commercialization, National Academic Press, Washington D.C.

NYSERDA (1998) (New York State Energy Research and Development Authority); Commerzialing Biomass Technologies in New York State. Producing a High-Value Chemical from Biomass (Levulinic acid). Project paper. [NYSERDA, Albany, New York]

Ohshima M, Proydak NI, Nishino N (1997). Effects of addition of lactic acid bacteria or previously fermented juice on the yield and nutritive value of alfalfa leaf protein concentrate by anaerobic fermentation. Anim Sci Tech 68:820-826.

Olson E S, Micelle R K, Schlag A J; Sharma R K (2001) Levulinate Esters from Biomass Wastes, ACS, Symposium Series 794, 51-63.

Ostrowski-Meissner HT (1983a). Protein concentrates from pasture herbage and their fractionation into feed- and food-grade products. In Leaf protein concentrates (Eds. Telek L, Graham HD). AVI Publishing: Westport, CT, USA, 437-489.

Ostrowski-Meissner, HT (1983b). Optimization of the protein extraction process from grasslands in temperate and subtropical regions. In Leaf protein concentrates (Eds. Telek L, Graham HD). AVI Publishing: Westport, CT, USA, 562-600.

Payene M J et al. (1993); Lectin-magnetic separation can enhance methods for the detection of Staphylococus aureus, Salmonelle enterididis and Listera monocytogenes. Food Microbiology, 10 75-83.

Pedone S, Selvaggini R, Fantozzi P (1995). Leaf protein availability in food: significance of the binding of phenolic compounds to ribulose-1,5-diphosphate carboxylase. Food Sci Technol-Leb 28:625-634.

Peitgen H O, Richter P H (1986) *The beauty of fractals*, Berlin.

Perl (1967) (ed.), *The Chemistry of Lignin* [Dekker, New York].

Pirie N W (1975) Leaf protein: a beneficiary of tribulation. *Nature*, 253 239-241.

Pirie NW (1987). *Leaf protein and its by-products in human and animal nutrition.* 2nd ed. Publisher Cambridge Univ. Press, Cambridge, Engl.

Pirie NW (1994). The bulk extraction and quality of leaf protein. In *Modern methods of plant analysis, Vol. 16 Vegetables and vegetables products* (Eds. Linskens HF, Jackson JF), Springer Verlag Berlin Heidelberg, 1-22.

Pirie N W, 1971 Leaf protein: In: *Agronomy, Preparation, Quality and Use.* In: IPB Handbook 20 [Blackwell Scientific Publications, Oxford,].

Prevot-D'Alvise N, Lesueur-Lambert C, Fertin-Bazus A, Fertin B, Dhulster P (2003). Development of a pilot process for the production of alfalfa peptide isolate. *J Chem Technol Biotechnol* 78:518-528.

Pusztai A, Bardocz S (1995) Physiological Role(s) of lectins in Plants and the Effects of their Inclusion in the Diet on the GUT and Metabolism of Mammals. In: *Phytochemical and Health.* [D.L. Gustine; H.E. Flores (ed.) American Society of Plant Physiologists].

Robowsky K D (1998) The potential of Amino acids from green plants. In: *The Green Biorefinery, concept of technology.* (First international symposium on green biorefinery) Neuruppin, Society of Ecological Technology and System Analysis, Berlin, ISBN 3-929672-06-5] pp.89-92.

Rouelle HM (1773). Sur les fécules ou parties vertes des plantes & sur la matière glutineuse ou végéto-animale. *J Méd Chir Pharm* 40:59.

Schertz F M (1938) Isolation of Chlorophyll, Carotene and Xanthophyll by Improved Methods. *Industrial and Engineering Chemistry*, 30,1073-75.

Schröder, M.R. et al (1993); Colosiation of Barley Lectin and Sporamin in Vacuolos Transgenic Tobaco Plants. *Plant Phys.,* 101 451-458.

Schwenke K D (1985) *Eiweisquellen der Zukunft* [Aulis-Verlag Deubner, Köln, 1985, ISBN 3-7614-0858-7] pp. 82.

Shearon W H, Gee O F (1949); Carotene and Chlorophyll. *Industrial and Engineering Chemistry*, 41, 220-26.

Sheen SJ (1991). Effect of succinylation on molecular and functional properties of soluble tobacco leaf protein *J Agr Food Chem* 39:1070-1074.

Sheen SJ, Sheen VL (1987). Characteristics of fraction-1-protein degradation by chemical and enzymatic treatments. *J Agr Food Chem* 35:948-952.

Sixth Symposium on Renewable Resources and Fourth European Symposium on Industrial Crops & Products (1999), Schriftenreihe Nachwachsende Rohstoffe, Vol. 14, Landwirtschaftsverlag, Münster (*a*) S. Girardeau, J. Aburto, C. Vaca-Garcia, I. Alric, E. Berredon, A *performing method of transesterfication of cellulose and amylose*; 832-836; (*b*) G. A. van Ingen, *Plastics based on proteins*, 837-839, (*c*) L. Heier, M. Heintges, K.-H. Kromer, *Short flaxfibre –production and quality*, 842-849.

Starke I, Holzberger A, Kamm B, Kleinpeter E. (2000) Qualitative and quantitative analysis of carbohydrates in green juices (wild mix grass and alfalfa) from a green biorefinery by gas chromatography/ mass spectrometry, *Fresenius J. Anal. Chem.*, 367, 65-72.

State of Brandenburg (2003), *Agriculture and food economy report (germ.)* [MELF Brandenburg (ed.), Potsdam].

Telek L, Graham HD (Eds.) (1983). *Leaf protein concentrates,* AIV Publishing: Westport, CT, USA.

Thomsen M H, Bech D, Kiel P (2004);Manufacturing of Stabilised Brown Juice for L-lysine production – from University Lab Scale over Pilot Scale to Industrial Production, *Chem Biochem Eng Q* 18 (1) 37-46.

Ulrich F, Tullius S G, Gärtner, M, Kamm B, Müller C, Neuhaus P, Steinmüller Th (2001) Mikroenkapsulierung von humanem Nebenschilddrüsengewebe mit Natriumcellulosesulfat und PolyDADMAC, *Acta Chir. Austriaca*, Vol. 33 (179) 15.

Verband der Chemischen Industrie e.V., (1994) *Fachvereinigung Organische Chemie, Konzeption des Fachausschusses Nachwachsende Rohstoffe.*

Wang JC, Kinsella JE (1976a). Functional properties of novel proteins: alfalfa leaf protein. *J Food Sci* 41:286-292.

Wang JC, Kinsella JE (1976b). Functional properties of alfalfa leaf protein: foaming. *J Food Sci* 41:498-501.

Wantanabe T, Fujishima A, Honda K, (1983) Dye-sensitive electrodes. In: *Energy Resour. Photochem. Cat.* [M.Grätzel (ed.), Academic Press] 359

Yang Y, Marczak ED, Usui H, Kawamura Y, Yoshikawa, M (2004). Antihypertensive properties of spinach leaf protein digests. *J Agr Food Chem* 52:2223-2225.

Zoebelin H (2001) *Dictionary of Renewable Resources,* WILEY-VCH, Weinheim

In: Progress in Biomass and Bioenergy Research
Editor: Steven F. Warnmer, pp. 33-52

ISBN: 1-60021-328-6
© 2007 Nova Science Publishers, Inc.

Chapter 2

EXPERIMENTAL ANALYSIS OF SMALL COMBUSTION THERMAL SYSTEMS BASED ON PELLETS

J.C. Morán[*1], *J.L. Míguez, E. Granada and J. Porteiro*
Universidad de Vigo - E.T.S. Ingenieros Industriales.
Lagoas-Marcosende, s/n. 36200 -Vigo (Pontevedra). Spain.

ABSTRACT

In this chapter a set joint of experimental techniques for assessing biomass combustion devices is presented. Small scale energy converters such as chimneys, boilers, stoves, etc, producing heat and/or hot water by combustion of biomass (wood, pellets, briquettes, etc.) are especially suited to domestic purposes. However, in regular commercial combustion conditions, this kind of use still has some disadvantages: besides the fact that some emissions (volatile organic carbons, carbon monoxide or NOx) may still be high, it is difficult to compare the quality and performance of equipment working in very different combustion conditions.

Due to their relatively low cost and the complexity of combustion in such devices, modelling by numerical analysis is seldom attempted. Controlling operational factors are usually designed and regulated based on the manufacturer's experience or on handbook values. In order to protect customers, and to assure compliance with minimum requirements for energy performance and maximum limits on pollutant emissions, several national and international regulations have been developed in recent years. Experimental analysis of these devices is a key technique for control and improvement.

1. INTRODUCTION

There are numerous discussions in the literature about the uncertainties in the analysis and design of thermal systems, taking into account their stochastic nature. This chapter presents a description of the problem in terms of experimental design and subsequent analysis. Although there are also several practical methods for performing relative analysis

---
[*] Universidad de Vigo - E.T.S. Ingenieros Industriales. Lagoas-Marcosende, s/n. 36200 -Vigo (Pontevedra). Spain.
e-mail: (1) jmoran@uvigo.es

the authors present here a still relatively unknown technique based on Grey Theory. Particularly, a method based on this theory called the grey relative analysis is a very simple, reliable analysis tool for time sequence and dispersed series. The use of grey relational analysis combined and compared with classical statistical experimental design and analysis applied to the optimising of a pellet combustion device is described as an application of the combination of the techniques.

The use of biomass, a renewable energy source, instead of fossil fuels in heat and power generation is increasing, and is a way of reducing global CO_2 and sulphur emissions and saving on diminishing fossil fuels. The use of wood pellets in small-scale systems is an increasing alternative for producing heat and hot water in the household sector. In Spain, just over half energy from renewable sources used involves biomass, which is limited almost solely to heat applications. With regard to thermal applications of biomass, the Renewable Energy Promotion Plan in Spain 1999-2010 [1] developed under the recommendations of the European White Paper [2] highlights the need for improvement in domestic equipment (typical residential equipment in a power range of 8 to 30 kW). The regional distribution of consumption is heavily related to the presence of the paper and pulp manufacturing sector, timber and food industries and the household sector. According to estimates from the 3rd National Forest Inventory, the amount of biomass in Galicia (northwest of Spain) exceeds 600,000 tm a year [3]. Residues derived directly from forest waste represented about 40% of the total (of which: Pine 38%, Eucalyptus Branches 22%, Crust of Eucalyptus 28%, Others 12%).

The energy use of residual wood is limited because of its low density. Waste from primary and secondary timber processing accounts for 20-30% in volume of all the raw material used. Biomass pelletising is a densification process that improves its handling characteristics, enhances its calorific value per volume, reduces transportation costs and produces a uniform fuel. Therefore, wood pellets are an alternative not only to non-treated wood but also to the traditional fossil fuels used in small-scale house heating systems [4, 5]. Typical characteristics of lignocellulosic pellets are low moisture content (<10% Wet Basis (WB)), high density (>700 $kg \cdot m^{-3}$) and a heating value around 17000 kJ/kg with a diameter between 5 and 12 mm [6]. The previous mentioned specifications make pellets very interesting for ordinary central heating systems. Moreover, they are an alternative to fossil fuels in small and average size boilers, and also in district heating plants and small scale house heating systems [7].

However, this kind of use still has some drawbacks, as emissions of pollutants such as VOC (Volatile Organic Carbons), CO (carbon monoxide) and NO_x still tend to be high (Table 1) [8]. Consequently, small-scale biomass pellet combustion technologies need further research to help to develop new equipment to meet current and future demand [9,10, 11].

Pellets burn cleaner than wood because the feeding rate is regulated and coupled with the accurate amount of air in order to obtain an optimum burn rate. Pellet fuel is made mainly of sawdust and shavings, or fine branches and leaves left over after processing trees for lumber and other wood products. Wood pellet handling is well-known and its aptness for boilers is one of its main advantages, especially in small-scale residence heating systems. In such pellet combustion devices, it is important to select appropriate operational parameters to achieve optimal performance. The desired parameters for solid fuels are usually determined on the basis of manufacturer experience or on handbook values [12]. But even with optimal operation parameters, it is difficult to compare the quality and performance of equipment working in such rather different combustion conditions [13, 14, 15]. The main reasons for high pollutant emissions are the relatively low combustion temperatures and the insufficient mixing rate between air and combustible gases resulting from pyrolysis and heterogeneous combustion of solid particles [16].

Table 1. Typical stove emissions

Emissions (mg/kWh)	Fuel	Gas	Pellet
CO	10	150	250
SO2	350	20	20
NOx	350	150	350
Particles	20	0	150
NMVOC	5	2	10

2. THEORY

One important issue would be to control the operational factors to achieve an overall optimal response, or at least some sort of indicator to quantify the gap related to the most possible favourable value. Thus, this kind of problem requires innovative management approaches, such as those that will ensure joint consideration of economic, ecological and social issues.

Another growing concern is consumer interest in environmental labels and certificates. Several methods and certification systems exist in Europe [17, 18, 19]. One of the most popular is the Swedish method called P-marking or labelling. From this point of view, integrating and qualifying all performance aspects of a device in a 0 to 1 scale is a very simple and intuitive labelling possibility.

Grey relational analysis is part of grey system theory, which is suitable for solving the complicated interrelationships between multiple factors and variables; in this case, how efficiency, emission variables and fuel prices clutch on to control variables. A grey system is defined as a system containing unknown information presented by grey numbers and grey variables [20]. In grey theory, black represents a system devoid of information, while white stands for complete information. Thus, the information that is either incomplete or undetermined is called grey. System theory was initiated by Deng in 1982, and has been widely applied in the last decade in China in several research fields [21, 22]. In many cases the improvement of one performance characteristic may be to the detriment of another.

In this research, the successively combination of experimental design classical statistic analysis and the use of grey relational analysis applied to the energetic and economic assessment and optimisation of combustion of wood pellets has been studied [23, 24, 25, 26, 27].

2.1. Experimental Design

In many situations it is desirable to learn something about the association between two attributes of an individual, material, a product, or for instance a complex process like pellet combustion. The problem then consists of determining the nature and degree of the relation. In regression analysis a regression equation is fitted to an observed point cloud of measured variables [28].

The effect caused by a simple variable when many influencing variables change at the same time is not easily to determine [29, 30]. Experimental design and subsequent statistic analysis is useful to find these correlations with minimum test runs and avoiding lost of generality in conclusions [31, 32, 33]. Statistics experimental design demands a minimum number of experiments avoiding a lost of generality in conclusions. These techniques provide valid conclusions and coupled with Anova define which are the most influential factors.

This work is focused on the analysis of the response of certain variables as an empirical function of one or more quantitative factors. A general form of this type of response function is

$$y = f(x_1, x_2, x_3 \ldots\ldots) \tag{1}$$

where y variable is the response, and x_1, x_2, x_3 are quantitative levels of the factors of interest. Knowledge of the form of the function f, will be found by fitting accurate correlations models to data obtained from designed experiments. When the response is a function of one or two factors, the fitted model is referred as a response surface because it can be graphed in one or two dimensions respectively, and can be explored to determine important characteristics such as optimum operating conditions or relevant tradeoffs.

Complete factorial designs require a lot of experiences depending on number of factors, k, and their corresponding number of levels, number of tests points in which the range of a selected variable will be measured.

$$\text{test runs} = n = p^k \tag{2}$$

In order to avoid an unaffordable number of experiences, three level factorial experiments are often conducted in order to fit such response surfaces. Two level factorial would require less tests, but could not detect curvature on the surface (maximums or minimums). As the number of factors increases, even three levels complete factorials become inefficient and impractical. Further, these design do not give equal precision for fitted responses at points (factor level combinations) that are at equal distances from the centre of the factor space, so that the exploration of the response surface would be dependent on the orientation of the design. In this study every experiment set has at least three levels in order to study possible second grade tendencies.

A design that has the property of equal precision at equal distances is termed a rotatable design. Rotability is a desirable property for response surfaces models because prior to the collection of data and the fitting of the response surface, the orientation of the design with respect to the surface is unknown. In general rotatable designs can be constructed from equally spaced points on circles or spheres. For more than two factors, design points should lie on a sphere, or a hyper sphere in four or more dimensions. The design points must also form a regular geometric figure. All 2^k complete factorials are rotatable but 3^k are not.

Table 2. Experimental layout examples in coded variables

Run Test	3 factors Central composite design			2 factors Complete factorial design	
	Factor a	Factor b	Factor c	Factor a	Factor b
1	-1	-1	-1	-1	-1
2	-1	-1	+1	-1	0
3	-1	+1	-1	-1	+1
4	-1	+1	+1	0	-1
5	+1	-1	-1	0	0
6	+1	-1	+1	0	+1
7	+1	+1	-1	+1	-1
8	-1.68	+1	+1	+1	0
9	+1.68	0	0	+1	+1
10	0	0	0		
11	0	-1.68	0		
12	0	+1.68	0		
13	0	0	-1.68		
14	0	0	+1.68		
15	0	0	0		
16	0	0	0		
17	0	0	0		

There are classes of designs for two or more factors that can be used in place of 3^k factorials for fitting second order polynomials to response surfaces. One design that makes more efficient use of the test runs is the central composite design, and an extra advantage is, that this design can be made rotatable. The total number of test runs in a central composite design, is based on a complete 2^k factorial design, but they only require enough extra observations to estimate the second order effects of the response surface. In this case

$$n = 2^k + 2k + m \tag{3}$$

which is less than 3^k, so that fewer observations are required than in a 3^k factorial design. "m" is the number of repeated observations at the design centre necessary to ignore interaction effects and "a" is the absolute extreme value of the desired studied interval. The central composite design can be made to be rotatable by choosing $a = F^{1/4}$, where F is the number of factorial points. ($F = 2^k$) In this case five levels of each factor has to be considered, coded (+a, +1,0,-1, -a)

2.2. Contrast of Equations

Many times it is difficult to decide which regression or correlation equation is appropriate. In the natural sciences every equation connects only precise quantities and the question of which variable is independent is often irrelevant.

But in our case the aim is to find cause, find out which are the actual factors that influence each and every variable. A measure for how well the regression line explains the observed values is the so called coefficient of determination R^2. How ever the factual interpretation of any statistical relations, and their testing for possible causal relation, lies outside the scope of statistical methodology. Even when stochastic dependence appears certain, one must bear in mind that the existence of a functional relation says nothing about a causal relation. The recognition of a causal correlation thus follows from the exclusion of other possibilities. Consequently the size of the correlation coefficient will not make any difference in this context. In order to contrast a correlation (mathematical model of the process) an evaluation process is necessary.

One common procedure is by the analysis of variance (ANOVA). As result correlations are evaluated and contrasted by ANOVA and validated considering phenomenon like multicollinearity (variance inflation factors, eigenvalues, …) influence factors (Cook distance…), auto-correlation (Durbin Watson…), heterocedasticity, etc.

2.3. Grey Theory

The grey system theory, first proposed by Professor Julong Deng in 1982 avoids he inherent defects of conventional statistical methods and only requires a limited amount of data to estimate the behaviour of an uncertain system. The major advantage of Grey theory is that it can handle both incomplete information and unclear problems very precisely. It serves as an analysis tool especially in cases when there is insufficient data [34]. It was recognized that the Grey relational analysis in Grey theory had been largely applied to project selection, prediction analysis, performance evaluation, and factor effect evaluation due to grey relational analysis software development.

There are numerous literature discussions regarding the uncertain in the analysis and design of the thermal system taking into account the stochastic nature. There are also several practical methods to perform relative analysis: parameter analytical method in mathematical statistics, orthogonal test analytical method and grey relative analytical method. Of these, the method of grey relative analysis is best suited to the relative analysis of time sequence and dispersed series [35, 36].

Considering a grey system where a set of variables y_i depend on certain independent factors x_j, all the possible combination of normalized experimental results x_{ij} can be expressed if the performance characteristic is the higher the better as:

$$x_{ij} = \frac{y_{ij} - \min_{j} y_{ij}}{\max_{j} y_{ij} - \min_{j} y_{ij}} \tag{4}$$

where y_{ij} is the ith experimental result in the jth experiment. Or if the performance characteristic is the lower the better pattern, as:

$$x_{ij} = \frac{y_{ij} - \max\limits_{j} y_{ij}}{\min\limits_{j} y_{ij} - \max\limits_{j} y_{ij}} \quad (5)$$

The grey relational coefficient, ξ_{ij}, is calculated to express the relationship between the ideal (best) and the actual normalized experimental results

$$\xi_{ij} = \frac{\min\limits_{i}\min\limits_{j}\left|x_i^0 - x_{ij}\right| + \zeta \max\limits_{i}\max\limits_{j}\left|x_i^0 - x_{ij}\right|}{\left|x_i^0 - x_{ij}\right| + \zeta \max\limits_{i}\max\limits_{j}\left|x_i^0 - x_{ij}\right|} \quad (6)$$

where x_i^0 is the ideal standardised result for the ith performance characteristic and ζ is the so-called distinguishing coefficient ($0 \leq \zeta \leq 1$).

The overall evaluation of the multiple variables is based on the grey relational grade, which is computed by averaging the grey relational coefficient corresponding to each performance characteristic,

$$\gamma_j = \frac{1}{m}\sum_{i=1}^{m}\xi_{ij} \quad (7)$$

where γ_j is the grey relational coefficient for the jth experiment and m is the number of performance characteristics (variables).

2.4. Procedure for the Experiment

To reduce the influence of any factors other than those studied here, a strict measurement procedure is applied. A stove of this kind never attains complete steady-state operation, mainly due to its batch feeding system, which in our case is controlled "pellet by pellet". If the measuring system is fast enough, an appreciable fluctuation may be detected after each pellet falls into the bed, as a result of the stir produced in the particularly thin bed that exists in this kind of fixed bed combustion system (this fluctuation is normally only detected by the CO analyser). For that reason we chose to follow the remaining variables (20 temperature readings, 6 pressure readings, 5 flows and the concentration of the rest of the gas species) until no variation was observed (which in several cases took more than 2 hours). Once the steady state was achieved, data recording was started. This took 20 minutes initially, but was subsequently extended to 60 minutes [37, 38].

3. EXPERIMENTAL

The University of Vigo (Spain) has developed a pilot plant to study pellet combustion in its different configurations and load conditions, taking into consideration both the energy and environmental viewpoints. The core of the plant is a commercial stove which has been modified to allow multiple controlled combustion conditions [39].

As with every complex system, in pellet combustion the effect caused by a single variable is not easily determined when many influential variables change at the same time. Experimental design techniques and subsequent statistical analysis of their results are useful to determine these correlations with a minimum number of test runs with no loss of generality in the conclusions. Many papers on this methodology have been published in many research fields since Box and Wilson developed the idea, some of which have dealt with combustion systems, but none has been reported that applies to pellet combustion in small systems.

Previous tests at our experimental plant revealed that the use of flue gases to preheat primary air improves the system's overall performance. Consequently the use of secondary air and/or gas recirculation, which has been successfully applied in bigger systems, may also be interesting [40].

The planning of the experimental part demands that the number of experiments should be minimised without loss of generality in conclusions. Due to the heterogeneity of the pellet, a minimum number of experiments is needed [41]. The most appropriate type of experiment and also the number of experiments are determined. These experiment design techniques provide valid conclusions and define which the most influential factors are. They also show the predicted tendencies in order to achieve a better efficiency of the general process [42].

3.1. Plant Description

An experimental plant has been used for this research (Figure 1). The main parts of this plant are the feeding system and the combustion unit.

Figure 1. Pilot Plant General View.

In order to study the combustion of different mixtures of pellets and even potentially cheap forest residues, a new feeding hopper for domestic biomass boilers has been implemented in the plant. This new hopper consists of a 200 mm wide (100 mm per product) band that drags both materials, pellet or forest residues, at the same time. The feeding rate can be regulated by means of a floodgate (0-40 mm height) or by changing the conveyor speed.

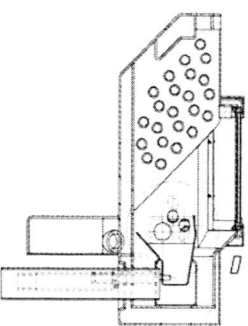

Figure 2. Combustion chamber scheme.

The combustion unit can be described as a pellet stove surrounded by a water jacket and vacuum operated, where pellets drop from the hopper into the combustion grate while combustion air is supplied upwards, through the grate. The stove has also a window door for ornamental purposes, maintenance and radiation heating purposes, which is closed for normal heat production. The combustion chamber has been adapted to allow multiple combinations of gas supply (primary and secondary air or already burnt gases, as desired) and also several access ports have been installed in the sidewalls and in all its gas-lines to facilitate the measurement of pressure, temperature, mass flow and gas concentration. The whole combustion device is isolated to calculate energy balance and efficiency.

Control and Analyses Equipment

- Pressure: is measured by SIEMENS transducers (Sitrans – P)
- Temperature: measured through thermocouples and RTD's (Pt-100).
- Combustion chamber exchangers, air and gas tubes have been isolated with polystyrene.
- Pellet flow rate control: moved by a controlled motor that determines the amount of pellet introduced in the combustion chamber. Flow is determinate measuring hopper weigh changes with an extensiometric gauge.
- Air flow rate: is controlled pressure drop by means three speed regulated fans.
- Water jacket temperature: is fixed with a regulated secondary water flow circuit
- Gas composition: gas analyses were carried out using a Horiba PG250 and TESTO-350 to measure gas concentrations of the main components involved: $NOx/SO_2/CO/CO_2/O_2$

3.2. Material

Typical characteristics of lignocelluloses pellets are low moisture content (<10% Wet Basis (WB)), high density (>700 kg·m^{-3}) and a heating value around 17000 kJ/kg with a diameter between 5 and 12 mm, and prices in Galicia are between 25 to 30 €/MWh. The previous mentioned specifications make pellets very interesting for ordinary central heating systems, and especially for single-family houses, which are very common in Galicia.

In this work, two relatively different kinds of pellet are tested.

Table 3. Pellets properties

	Pellet 1	Pellet 2
Diameter (mm)	7	6.6
Length (mm)	7-21	10-15
Density[1] (kg·m^{-3})	1166	943.4
Moisture (% WB)	8.5	9.5
Proximate Analysis (wet % dry basis)		
Ash (550 °C)	0.68	0.90
Ultimate Analysis (wet % dry basis Ash free)		
Carbon	51.7	52.0
Hydrogen	6.7	5.3
Oxygen	40.7	42.0
Nitrogen	0.17	0.92
Sulphur	< 0.05	< 0.05

[1] Geometric method.

4. RESULTS AND DISCUSSION

Previous test analysis shows the dominance of just two main factors m_p and n. To study secondary air influence an amount M_2 (air flow rate in l/s) of the total air introduced into the stove, $M_{T\,A}$ (air flow rate in l/s), is forced through the lateral pipes. The ratio sa = M_2 / $M_{T\,A}$ is selected as the factor for Set SA. The study covers a range from 0 to 0.30 of this factor.

Analogously, to study the effect of smoke recirculation part of the flue gas, M_R (gas flow rate in l/s), of the combustion chamber M_{TS} (gas flow rate in l/s) is again introduced into the stove, sr = M_R / $M_{T\,S}$ is the third factor selected for Set SR. The study covers a range from 0 to 0.30.

Changes in the factors have an effect on variables but there is no simple direct relation in these changes, as each variable also depends on other factors. Therefore, correlations between factors and variables needed a statistical approach.

4.1. Variables and Factors

In accordance with previous tests, two different experimental designs are used to determine the controlling factors for pellet combustion in such stoves. The variables studied for different combustion conditions are heat in water Q_w , heat in flue gases Q_s , stove efficiency η_{st} , combustion efficiency η_c and exhaust gas composition. Statistical experimental design coupled with ANOVA defines which the most influential factors are.

Table 4. Factors and variables

Factors	
n	Stoichiometric
m_p (gr/s)	Pellet supply
sa	Secondary air
sr	Recirculation
Variables	
Q_W (kW)	Heat in water
Q_S (kW)	Heat in flue
Q_T (kW) =	Liberated heat
Q_A (kW) =	Supplied heat
η_{st} = Q_W /	Stove
$\eta_c = Q_T / Q_A$	Combustion
CO (g/kWh	CO emissions
NO (g/kWh	NO emissions

[1]LHV = low heating value wet basis

Table 5. SET SA and SET SR factor levels before randomising

	Set SA			Set SR		
Coded levels	m_p(gr/s)	n	sa (%)	m_p (gr/s)	n	sr (%)
+1.68	2.0	2.00	30	2.0	2.00	30
+1	1.8	1.84	24	1.8	1.84	24
0	1.5	1.60	15	1.5	1.60	15
-1	1.2	1.36	6	1.2	1.36	6
-1.68	1.0	1.20	0	1.0	1.20	0

4.1.1. Secondary Air (Set Sa)

The heat transferred to water depends on the amount of pellet supplied (m_p), and, to a lesser extent, on n. The amount of secondary air used has no significant influence. Heat in smoke increases linearly with the contribution of both m_p and n. Secondary air reduces CO emissions, with n still being the critical factor, but the amount of secondary air is also significant even for combustion conditions where CO emissions are at their lowest (n approximately 1.65).

Table 6. Experimental results. Set SA

Test	Mp g/s	n	sa	η_{st}	η_c	Qw	CO g/kWh	NO g/ kWh
SA -1	2.00	1.60	0.15	0.73	0.87	23.25	17.26	0.63
SA - 2	1.50	1.60	0.30	0.75	0.86	19.97	**2.72**	0.68
SA - 3	1.80	1.80	0.24	0.72	0.87	21.92	2.92	0.74
SA - 4	1.20	1.40	0.06	0.74	0.81	15.31	40.56	0.45
SA - 5	1.20	1.40	0.24	0.74	0.81	15.15	21.29	0.51
SA - 6	1.20	1.80	0.06	0.77	0.87	15.84	9.90	0.61
SA - 7	1.20	1.80	0.24	0.78	**0.89**	16.30	3.67	0.64
SA - 8	1.50	1.60	0.15	0.71	0.82	18.66	23.71	0.53
SA - 9	1.50	1.20	0.15	0.62	0.69	16.21	40.45	**0.43**
SA - 10	1.50	2.00	0.15	0.74	0.88	19.07	4.87	0.67
SA - 11	1.50	1.60	0.15	0.73	0.84	18.65	15.20	0.47
SA - 12	1.00	1.60	0.15	**0.80**	0.88	12.99	11.12	0.52
SA - 13	1.50	1.60	0.00	0.74	0.85	18.99	17.89	0.53
SA - 14	1.80	1.40	0.24	0.70	0.81	21.21	18.25	0.49
SA - 15	1.80	1.80	0.06	0.68	0.83	21.56	3.55	0.85
SA - 16	1.80	1.40	0.06	0.70	0.81	20.9	23.49	0.45
SA - 17	1.50	1.60	0.15	0.74	0.85	18.6	12.32	0.48
Lowest value				0.62	0.69	12.99	**2.72**	**0.43**
Highest value				**0.80**	**0.89**	**23.25**	40.56	0.85

4.1.2. Smoke Recirculation (Set SR)

Table 7. Experimental results. Set SR

Test	Mp g/s	n	sr	η_{st}	η_c	Qw	CO g/kWh	NO g/ kWh
SR -1	1.2	1.36	0.06	0.81	0.89	17.46	23.84	0.25
SR - 2	1.5	1.20	0.15	0.70	0.79	19.45	27.86	**0.21**
SR - 3	1.5	1.60	0.15	0.80	0.93	21.40	7.04	0.33
SR - 4	1.2	1.84	0.06	0.82	0.94	17.85	3.45	0.34
SR - 5	1.5	1.60	0.30	0.84	**0.99**	20.11	2.41	0.40
SR - 6	1.8	1.36	0.06	0.80	0.92	23.81	35.31	0.30
SR - 7	1.5	2.00	0.15	0.73	0.88	20.39	2.90	0.42
SR - 8	1.8	1.84	0.06	0.79	0.95	24.26	4.80	0.42
SR - 9	1.0	1.60	0.15	**0.87**	0.95	15.49	3.02	0.27
SR - 10	1.2	1.36	0.24	0.83	0.91	17.33	7.05	0.25
SR - 11	2.0	1.60	0.15	0.76	0.91	**25.02**	3.04	0.39
SR - 12	1.5	1.60	0.15	0.78	0.91	20.47	5.76	0.33
SR - 13	1.8	1.36	0.24	0.72	0.84	23.08	2.37	0.32
SR - 14	1.8	1.84	0.24	0.77	0.94	22.39	1.86	0.42
SR - 15	1.5	1.60	0.00	**0.87**	**0.99**	21.50	20.99	0.33
SR - 16	1.5	1.60	0.15	0.83	0.95	21.72	7.21	0.32
SR - 17	1.2	1.84	0.24	0.80	0.92	17.39	**1.25**	0.35
Lowest value				0.70	0.79	15.49	**1.25**	**0.21**
Highest value				**0.87**	**0.99**	**25.02**	35.31	0.42

The heat transferred to water depends mainly on the two factors m_p and n, with m_p being the more significant of the two. The amount of recirculation has only a small influence. Heat in smoke (Q_S) increases linearly with the contribution of both m_p and n. Emissions, however, are affected by recirculation. CO can be reduced significantly by ensuring a proper air-fuel rate and with a relatively low degree of recirculation of smoke.

4.2. Statistical Analysis

An extensive statistical analysis of each variable may obtained contrasted correlations for both sets of measures. Relatively high correlation index R^2 is obtained in the regressions for both energy and emissions variables. Concerning energy efficiency neither recirculation of air nor smoke has major advantages, at least directly. Best efficiency is obtained with low loads and high air fuel ratio ($n = 2$). However if CO is an important question recirculation of smoke reduces remarkably the emissions. Greatest reduction is obtained with low air fuel ratios ($n < 1.5$). In addition, as NO emission increase with the air fuel ratio, recirculation of smoke can help to get an overall optimum combustion as it make possible to work with lower n, low NO emissions, without a big loss in efficiency (only 5%) and lower CO emission as combustion without recirculation and an air fuel ratio of $n = 2$.

Table 8. Statistical results

	R^2_C
SET SA	
$QW(kW) = 9.83 \cdot mp + 2.35 \cdot n$	99.9%
$QS(kW) = -5.62 + 3.39 \cdot mp + 2.14 \cdot n$	98.6%
CO gr/kWh$_{Pellet}$ = $62.68/n^2 - 62.43 \cdot (sa)$	84.1%
NO gr/kWh$_{Pellet}$ = $0.135 \cdot n^2 + 0.145 \cdot mp$	98.3%
$\eta_C = 0.78 + 0.26 \cdot n \cdot mp^{-1} - 0.33 \cdot mp^{-1}$	
$\eta st = 0.58 + 0.14 \cdot (n \cdot mp^{-1})$	
SET SR	
$QW(kW) = 10.9 \cdot mp + 2.58 \cdot n$	99.8%
$QS(kW) = -5.48 + 3.51 \cdot mp + 2.17 \cdot n$	98.6%
CO gr/kWh$_{Pellet}$ = $51.33/n^2 - 76.7 \cdot (sr)$	81.9%
NO gr/kWh$_{Pellet}$ = $0.07 \cdot n^2 + 0.10 \cdot mp$	99.5%
$\eta_C (\%) = 82.84 + 27.3 \cdot n \cdot mp^{-1} - 31.5 \cdot mp^{-1}$	
$\eta st (\%) = 62.67 + 14.8 \cdot (n \cdot mp^{-1})$	

The optimisation of pellet combustion as a whole means achieving that combination of independent process parameters (factors) that leads to the highest possible values of heat in water, with the best achievable efficiencies, and the lowest attainable emissions of NO and CO. As a result, an improvement of one variable may be in detriment to another variable that depends on the same controlling factor (example SET SA in Table 9).

Table 9. Performance characteristics. SET SA

Correlated Variable	Goal	Factors		
		m_p	n	Sa
Q_W	↑	↑	↑	Not coupled
η_{st}	↑	↓	↑	Not coupled
η_c	↑	↑	↑	Not coupled
CO	↓	Not coupled	↑	↓
NO	↓	↓	↓	Not coupled

Optimisation of the multiple variables cannot be as straight-forward as the optimisation of a single variable. Consequently the problem is not only to detect which is the best configuration but also to define the optimal combustion that would involve simultaneous maximal stove and combustion efficiency (highest value) and minimum emissions (lowest value). Thus, the optimal operation conditions cannot be easily established and need the definition of the grey relational grade (GRG). Through this new variable, tests can not only be compared, but also the optimisation studying the correlation of the GRG with the factors means the optimisation of the whole combustion process.

4.3. Grey Relational Grade

In the grey relational analysis method, experimental data (combustion efficiency, stove efficiency, CO emissions, NO emissions) are first normalised in the range between zero and one, which is also called the grey relational generation. Subsequently, the grey relational coefficient is calculated from the normalised experimental data to express the relationship between the desired and actual experimental data. The grey relational grade is then computed by averaging the grey relational coefficient corresponding to each process response. The overall evaluation of the multiple process responses is based on the grey relational grade.

Experimental results of selected performance characteristics (stove and combustion efficiency and CO and NO emissions) were standardised to calculate the grey relational coefficient of each variable. The grey relational grade (GRG) was calculated by averaging the grey relational coefficient. Certain experimental results are given in Table 10.

The influence of each factor using the averaged grey relational grade for each level was analysed. A higher value of the grey relational grade represents a stronger relational degree between the reference sequence $x0(k)$ and the given sequence $xi(k)$. As mentioned before, the reference sequence $x0(k)$ is the best process response in the experimental layout. In other words, optimisation of the complicated multiple process responses can be converted into optimisation of a single grey relational grade. Depending on which group of performance characteristics are included in order to calculate the Grey Relational Grade, different scales are obtained.

As the grey relational coefficient, ξ_{ij}, express the relationship between the ideal (best) and the actual normalized experimental results, so best operational conditions for η_{st} performance were obtained in tests coded SR-9 and SR-15, those for η_c performance were obtained in tests coded SR-5 and SR-15, those for CO performance in test number SR-17, and those for NO performance was obtained in test number SR-2.

Just as classical statistical analysis show, but straightforwardly, the grey relational grade including all tests (SET SA+ SET SR) illustrate that SET SR is better (Figure 3).

Table 10. Grey relational grade SET SA + SET SR

Test	mp	n	sa	sr	η_{st}	η_c	CO	NO	GRG
	g/s				ξ_{ij}	ξ_{ij}	ξ_{ij}	ξ_{ij}	
SA-1	2.0	1.6	0.15	-	0.48	0.56	0.55	0.43	0.50
SA-2	1.5	1.6	0.30	-	0.51	0.53	0.93	0.40	0.60
SA-3	1.8	1.8	0.24	-	0.45	0.55	0.92	0.37	0.57
SA-4	1.2	1.4	0.06	-	0.49	0.45	0.33	0.57	0.46
SA-5	1.2	1.4	0.24	-	0.49	0.46	0.50	0.52	0.49
SA-6	1.2	1.8	0.06	-	0.56	0.56	0.69	0.45	0.56
	1.2	1.8	0.24	-	0.59	0.60	0.89	0.43	0.63
SA-8	1.5	1.6	0.15	-	0.45	0.47	0.47	0.50	0.47
SA-9	1.5	1.2	0.15	-	0.33	0.33	0.33	0.60	0.40
SA-10	1.5	2.0	0.15	-	0.49	0.58	0.84	0.41	0.58
SA-11	1.5	1.6	0.15	-	0.47	0.50	0.58	0.55	0.53
SA-12	1.0	1.6	0.15	-	0.63	0.57	0.67	0.50	0.59
SA-13	1.5	1.6	0.00	-	0.50	0.52	0.54	0.50	0.51
SA-14	1.8	1.4	0.24	-	0.42	0.46	0.54	0.54	0.49
SA-15	1.8	1.8	0.06	-	0.40	0.48	0.90	0.33	0.53
SA-16	1.8	1.4	0.06	-	0.43	0.45	0.47	0.57	0.48
SA-17	1.5	1.6	0.15	-	0.49	0.51	0.64	0.54	0.55
SR -1	1.2	1.36	-	0.06	0.68	0.60	0.47	0.89	0.66
SR -2	1.5	1.20	-	0.15	0.43	0.43	0.42	**1**	0.57
SR -3	1.5	1.60	-	0.15	0.65	0.71	0.77	0.73	0.71
SR -4	1.2	1.84	-	0.06	0.72	0.75	0.90	0.71	0.77
SR -5	1.5	1.60	-	0.30	0.81	**1**	0.94	0.62	0.84
SR -6	1.8	1.36	-	0.06	0.65	0.68	0.37	0.79	0.62
SR -7	1.5	2.00	-	0.15	0.48	0.58	0.92	0.60	0.64
SR -8	1.8	1.84	-	0.06	0.61	0.79	0.85	0.60	0.71
SR -9	1.0	1.60	-	0.15	**1**	0.79	0.92	0.84	**0.89**
SR -10	1.2	1.36	-	0.24	0.76	0.65	0.77	0.91	0.77
SR -11	2.0	1.60	-	0.15	0.54	0.65	0.92	0.64	0.69
SR -12	1.5	1.60	-	0.15	0.59	0.65	0.81	0.73	0.70
SR -13	1.8	1.36	-	0.24	0.46	0.50	0.95	0.75	0.66
SR -14	1.8	1.84	-	0.24	0.56	0.75	0.97	0.60	0.72
SR -15	1.5	1.60	-	0.00	**1**	**1**	0.50	0.73	0.81
SR -16	1.5	1.60	-	0.15	0.76	0.79	0.77	0.74	0.77
SR -17	1.2	1.84	-	0.24	0.65	0.68	**1**	0.70	0.76

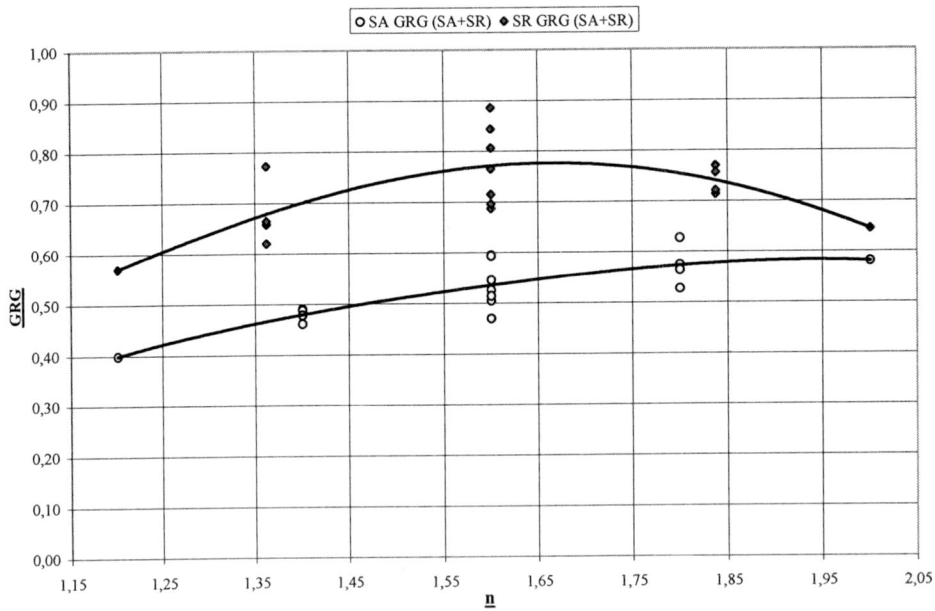

Figure 3. Grey relational grade. Configuration analysis.

The influence of each factor on grey relational grade can be studied as each factor is an independent variable.

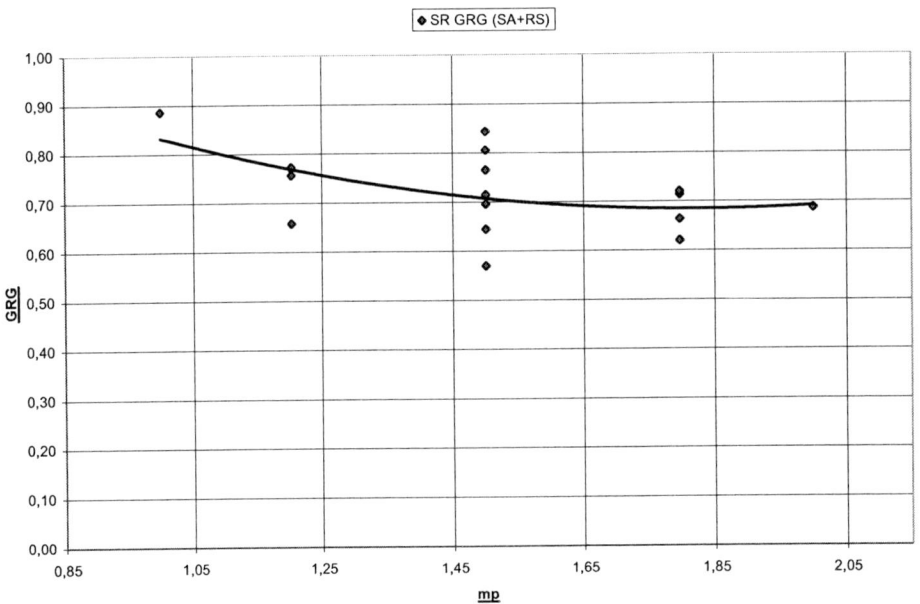

Figure 4. Grey relational grade. Factor dependence analysis.

According to this factor analysis, optimum is obtained around mp = 1.00, n = 1.60, sr = 30%. This optimum was verified through a confirmation experiment and recalculation of new relational coefficients and grey relational grades.

Table 11. GRG$_N$ - SET SA + SET SR + Optimum

Test	mp g/s	n	sa	sr	η_{st}	η_c	CO g/kWh	NO g/ kWh	GRG$_N$
SR -9	1.0	1.60	-	0.15	0.87	0.95	3.02	0.27	0.87
Optimum	2.0	1.6	-	-	0.88	0.97	1.23	0.36	0.89

The inclusion of confirmation test achieved a new highest GRG$_N$ on this new scale (SET SA + SET SR + OPTIMUM), higher than previous best value test (former value 0.89, in new scale 0.87). This optimum means that although not all performance characteristics achieved the best values individually, this combination of factors produced the best possible overall averaged values.

CONCLUSION

This work carried out relative analyses on the multivariable relationships of pellet combustion in an improved domestic pellet boiler by means of the grey relative analytical method. Grey relational analysis is part of a Grey theory and is here applied to evaluate the combustion process.

In addition to classical statistic techniques and by means of the grey rational grade, this methodology permits qualifying all the aspects of combustion with just a single variable. Consequently, the optimisation of complicated multiple responses can be converted into the optimisation of a normalized single grey relational grade. Indeed, the confirmation experiment shows that factor dependence analysis of the grey relational grade led to an "optimal combustion".

REFERENCES

[1]   Instituto para la Diversificación y Ahorro de la Energía, IDAE. *Plan de Fomento de las Energías Renovables en España.* 1999

[2]   The European Commission. 97- 599 Energy for the Future: Renewable Sources of Energy. *White Paper for a Community Strategy and Action Plan.* 1997

[3]   Hernandez, C . Biomass in Spain: Activities, policies and strategies. The renewable energy promotion plan. 1st World Conference on Biomass for Energy and Industry. Sevilla. London: James and James Science Publishers Ltd. 2000. pp 1213-1216.

[4]   OPET Sweden Framework. Refined biomass - a source for climate change and business opportunities Work package: Market development of refined biomass. *EM-BM-1.* 2001

[5]   Gustavsson, L; Tullin, C. and Wrande I. Small-scale biomass combustion in Sweden - research towards a sustainable society. Proceedings of the 1st World Conference on Biomass for Energy and Industry. Sevilla. London: James and James Science Publishers Ltd. 2000 pp 1553 - 1555.

[6]   Rösch, C.; Kaltschmitt, M. and Limbrick A. Standardisation of Solid Biofuels in Europe. Proceedings of the 1st World Conference on Biomass for Energy and Industry. Sevilla. London: James and James Science Publishers Ltd. 2000. pp 609-612.

[7]   Obernberger, I. and Thek G. The Current State of Austrian Pellet Boiler Technology. BIOS Bioenergiesysteme GmbH. First World Conference on Pellets. Stockholm. Sweden. 2002. pp 45-48.

[8]   Instituto para la Diversificación y Ahorro de la Energía, IDEA. Biocombustibles para edificios de viviendas.Ministerio de Economía. España M-49974-2002.

[9]   Cotton, R.A and Giffard, A. Introducing wood pellet fuel to the UK. Renewable Heat and Powder Ltd. DTI Sustainable Energy Programmes – Altener Programme. 2001

[10]   Frandsen, S. Biomass Survey in Europe. Country Report of Denmark. Danish Technological Institute. Danish Energy Authority. *European Bioenergy Networks.* 2002.

[11]   Egger, Ch. and Öhlinger, Ch. Strategies and Methods to Create a New Market. First World Conference on Pellets. Stockholm. Sweden. 2002. pp 35-37

[12]   Lasselsberger, L. Quality Marking and Environmental Testing of Small-Scale Biomass Boilers in Austria. *Landtechnische Forschung* (2000).

[13]   Skreiberg, O. and Saanum, O. Comparison of Emission Levels of Different Air Pollution Components from Various Biomass Combustion Installations in the IEA Countries. Supplemental Report to the IEA-Project -Emissions from Biomass Combustion. The Norwegian Institute of Technology. Institute of Thermal Energy and Hydropower. 1995.

[14]   Fiedler, F. The state of the art of small-scale pellet-based heating systems and relevant regulations in Sweden, Austria and Germany. *Renewable and Sustainable Energy Reviews.* 2004 vol, 8 pp 201 - 221.

[15]   Dias, J:; Costa, M. and Azevedo, J.L.T. Test of small domestic boiler using different pellets. First World Conference on Pellets. Stockholm. Sweden. 2002. pp 137-141

[16]   Sjöstrom E. Wood Chemistry: Fundamentals and Applications. Academic Press. 1993.

[17]   Oravainen, H. Testing methods and emission requirements for small boilers (<300kW) in Europe. *VTT Energy.* Helsinki - Finland.2000.

[18]   EN 303-5 Heating boilers. Part 5: Heating boilers for solid fuels, hand and automatically stocked, nominal heat output of up to 300 kW. *Terminology, requirements, testing and marking.*

[19]   EN-12809. 2001 Residential independent boilers fired by solid fuel - Nominal heat output up to 50 kW - *Requirements and test methods*

[20]   Wu, H. The Introduction of Grey Analysis. Gauli Publishing Co. *Taipei.* China. 1996.

[21]   Deng, J. Grey System:Society and Economy. Publishing House of National Defense Industry. *Wuhan.* China 1985.

[22]   Deng, J. Introduction to Grey System Theory. The Journal of Grey Systems. Sci-Tech Information Services, England and China Petroleum Industry Press, P. R. China. 1989. vol 1.

[23] Coetzer, R.L.J. and Keyser, M.J. Experimental design and statistical evaluation of a full scale gasification project. *Fuel Processing Technology.* 2003. vol 80 pp 263-278

[24] Lin, J.L. and Lin, C.L. The Use of the Orthogonal Array with Grey Relational Analysis to Optimize the Electrical Discharge Machining Process with Multiple Performance Characteristics. *International Journal of Machine Tools and Manufacture.* London. GB. 2002 vol 42

[25] Lin, C.L. and Lin, J.L. Optimisation of the EDM Process Based on the Orthogonal Array with Fuzzy Logic and Grey Relational Analysis Method. *The International Journal of Advanced Manufacturing.* London. GB. 2002 vol 42

[26] Huang, J.T and Liao, Y.S. Optimization of Machining Parameters of Wire-EDM Based on Grey Relational and Statistical Analyses. *International Journal of Production Research.* 2003. vol. 41- 8 pp 1707-1720.

[27] Huang, J.T and Liao, Y.S. Application of Grey Relational Analysis to Machining Parameters Determination of Wire Electrical Discharge Machining. Proceedings International Symposium for Electromachining, (ISEM XIII). Fundación Tekniker. Bilbao. Spain .2001

[28] Sánchez de Rivera, P. Modelos lineales y series temporales. Alianza Universidad Textos. Spain.1989

[29] Mason, R.L.; Gunst, R.F. and Hess, J.L. Statistical design and analysis of experiments.Willey and Sons.1989

[30] Lothar Sachs. Applied Statistics. A Handbook of Techniques. Second Edition. *Spinger Verlag.* 1984

[31] Nordin, A.; Eriksson L. and Öhman M. NO reduction in a fluidized bed combustor with primary measures and selective non-catalytic reduction: A screening study using statistical experimental designs *Fuel.* 1995. vol 74 pp 128-135.

[32] Box, G.E.P. and Wilson, K.B. On the experimental attainment of optimun conditions. *J. R. Stat. Soc. B.* 1951 vol 13 pp 1-45

[33] Yunyan G. Design Method of Orthogonal and Regression Test. Publishing House of Shanghai Traffic University. Shanghai. China. 1980.

[34] Kun-Li, W. Grey Systems. Modelling and Prediction. Yang's Scientific Press. 2004

[35] Xueming, W. and Jianjun, L. Simple Course of Grey System Method. Publishing House of Chengdu Technology College. Chengdu. China. 1993.

[36] Lin, Y.; Chen, M-y. and Liu, S. Theory of grey systems: capturing uncertainties of grey information. *Kybernetes.* Emerald Group Publishing Limited. 2004. vol 33-2 pp. 196-218.

[37] Míguez, J.L.; Porteiro, J.; Morán, J.C.; Granada, E. and López, L.M. Description of a pilot lignocellulosic pellets stove plant. Sixth European Conference on Industral Furnaces and Boilers. *Estoril.* Portugal. 2002

[38] AENOR.UNE 9-044-086. Spain. 1986

[39] Granada, E.; Morán, J.C., Porteiro, J.; Miguez, J.L. and Lareo, G.. Pilot lignocellulosic pellet stove plant. First World Conference on Pellets. Stockholm. Sweden. 2002. pp 147-151

[40] Houck, J. E.; Scott A. T.; Purvis, C.R. Kariher, P.H.; Crouch, J. and Van Buren, M.J. Low emission and high efficiency residential pellet-fired heaters. Proceedings of the Ninth Biennial Bioenergy Conference, Buffalo, NY -USA. 2000

[41] Morán, J.C.; Granada E.; Porteiro,J. and Míguez, J.L. Pellet Combustion in Stove: Performance and Emissions Statistical Approach. International Conference on Renewable Energy and Power Quality. *Vigo.* Spain. 2003. pp 122-126

[42] Morán, J.C.; Granada E.; Porteiro,J. and Míguez, J.L. Experimental modelling of a pilot lignocellulosic pellets stove plant. *Biomass and Bioenergy.* 2004. vol 27- 6 pp 577-583.

In: Progress in Biomass and Bioenergy Research
Editor: Steven F. Warnmer, pp. 53-100

ISBN: 1-60021-328-6

Chapter 3

NEGATIVE EMISSION BIOMASS TECHNOLOGIES IN AN UNCERTAIN CLIMATE FUTURE

Kenneth Möllersten[1], Zuzana Chladná[1,2],
Miroslav Chladný[3] and Michael Obersteiner[1]

[1]International Institute for Applied Systems Analysis,
Schlossplatz 1, A-2361 Laxenburg, Austria
[2]Department of Applied Mathematics and Statistcs,
Comenius University, 84248 Bratislava, Slovakia
[3] Department of Computer Science, Comenius University,
Bratislava, Slovakia

ABSTRACT

Mitigation of and adaptation to climate change belong to the most pressing global challenges for the 21st century. Major mitigation options include improved energy efficiency, shifting towards less carbon-intensive fossil fuels, increased use of energy sources with near-zero emissions, such as renewables and nuclear, CO_2 capture and permanent storage (CCS), and carbon sequestration by protection and enhancement of biological absorption capacity in forests and soils.

Bioenergy is one of several energy sources which could provide society with energy services with near-zero emissions. Bioenergy has a unique feature, however, which distinguishes it from other low-emitting energy supply options, such as solar, wind, nuclear, and clean fossil energy technologies. Bioenergy conversion could be integrated with a process which separates carbon. If the biomass feedstock is sustainably produced and the separated carbon is subsequently isolated from the atmosphere for a very long time the entire process becomes a continuous carbon sink – in other words such technologies yield negative CO_2 emissions. Negative emission biomass technologies can be centralised or distributed; Centralised negative emission biomass technologies, biomass energy with CO_2 capture and storage (BECS), build on the conversion of biomass into energy carriers in centralised conversion plants integrated with CO_2 capture. The captured CO_2 is subsequently transported and stored in geological formations.

Distributed negative emission biomass technologies are based on the production of long-term carbon-sequestering charcoal soil amendment, with or without co-production of biofuels.

In this chapter a BECS implementation scenario study is presented. The study analyses investments in BECS in a pulp and paper mill environment. The investment analysis is carried out within a real options framework taking into account the potential revenue from trading generated emission allowances on a carbon market. Uncertainty is considered in the economic modelling through the use of stochastically correlated price processes of one input price (biomass) and two output prices (electricity and CO_2 emission permits) that are consistent with shadow price trajectories of a large-scale global energy model. The results suggest that BECS can be economically feasible within approximately 40 years.

The chapter also discusses Research and Development needs for better understanding of the future overall potential of negative emission biomass technology implementation.

LIST OF ACRONYMS

ADt	Air-dry tonne
BIG	Biomass integrated gasification
BIG/CC	Biomass integrated gasifier with combined cycle
BECS	Biomass energy with CO_2 capture and storage
BLG	Black liquor gasification
BLG/CC	Black liquor integrated gasifier with combined cycle
CCS	CO_2 capture and storage
CHP	Combined heat and power production
COE	Cost of electricity
GHG	Greenhouse gas
HRSG	Heat recovery steam generator
IPCC	Intergovernmental panel on climate change
IPPM	Integrated pulp and paper mill
LBM	Limited biomass model
MPM	Market pulp mill
MWe	Megawatt electric
ppm	Parts per million
UNFCCC	United Nations framework convention on climate change

INTRODUCTION

Evidence is mounting that human-induced increase in greenhouse gas (GHG) levels in the atmosphere is creating one of the direst environmental changes in human history. Since the pre-industrial era, the concentration of CO_2 in the atmosphere has increased from 280 to over 370 parts per million by volume (ppmv), mainly as a result of the burning of fossil fuels and land clearing.

The third assessment report of the Intergovernmental Panel on Climate Change (IPCC) states that most of the observed global warming over the last 50 years is likely to have been due to the increase in greenhouse gas (GHG) concentrations in the atmosphere [IPCC, 2001a]. The IPCC further concludes that the stabilisation of the atmospheric CO_2 concentration requires CO_2 emissions to eventually drop well below current levels.

The required reduction in net emissions to the atmosphere can be realised by means of [IPCC, 2001b]:

- Demand reductions and/or effiency improvements
- Substitution among fossil fuels
- Switching to renewables or nuclear energy
- CO_2 capture and storage (CCS)
- Carbon sink enhancement.

The IPCC [2001b] concludes that none of these measures alone can achieve the emission reductions required. Thus, a portfolio of technologies needs to be developed and adopted. The choice of a particular stabilization level from any given baseline significantly affects the technology portfolio required for achieving the necessary emissions mitigation. Generally, a wider range of technological measures and their widespread diffusion and more intensive use is required for stabilizing at 450 ppmv compared with stabilization at higher levels [Nakicenovic et al., 2002].

Technology plays a crucial role in addressing the challenge of long-term GHG abatement. One way of categorising technologies is to arrange them in the following three groups:

Technologies that give rise to substantial net CO_2 emissions and thus contribute to a build-up of GHGs in the atmosphere; Conventional energy technologies based on fossil fuels is the most obvious example of technologies that belong to this category.

Technologies with near-zero net CO_2 emissions, i.e. technologies that can be regarded as more or less climate-neutral; Renewables-based technologies such as solar and geothermal energy, hydro power, wind power, and sustainable bioenergy[1] along with nuclear energy are well known examples of this category. Among possible energy technologies with near-zero emissions, fossil fuel utilisation in combination with so-called CO_2 capture and storage (CCS) is receiving increasing recognition as a mitigation option.[2] CCS is the separation of CO_2 from anthropogenic sources, transport to a storage location, and isolation from the atmosphere, see Fig. 1. Fossil fuel utilisation with CCS is seen by many as a key future technological option to reduce CO_2 emissions that can be attributed with a large global CO_2 reduction potential. Technologies for CO_2 capture and transportation are technically proven and commercially available, although there is a large potential for improvements both regarding efficiency and costs. It remains to be established, however, criteria and standards for safe CO_2 storage over very long time periods. Moreover estimates of regional and global

---

[1] Conventional bioenergy can be carbon-neutral or -positive depending on how land use is affected by the biomass source. If biomass extracted for energy purposes from an existing biomass stock is subsequently replanted, the same amount of carbon that is releases from combustion is extracted from the atmosphere through photosynthesis and the bioenergy system is generally CO_2 neutral (for a detailed discussion see, for example, Schlamadinger et al., 1997).

[2] See IPCC [2005] for a comprehensive overview.

storage potentials are still very uncertain (see, for example, Bradshaw et al., 2006). With respect to CCS, scientists, policy makers and industry have largely focussed on fossil fuel-based systems. However, it is important to note that capturing CO_2 from an emission source and storing it away from the atmosphere has the same climate change-mitigating effect regardless if the CO_2 capture is applied to emissions from fossil fuels or from biomass.

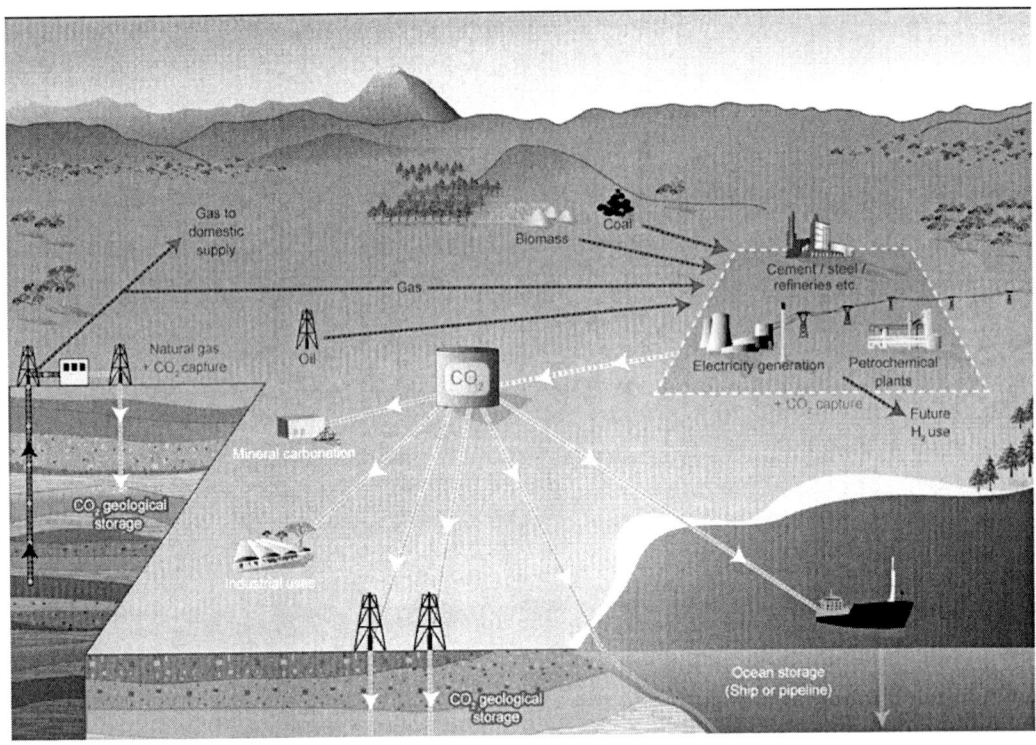

Figure 1. Schematic diagram of possible CCS systems (Source: IPCC, 2005).

Negative emission technologies, i.e. technologies that remove CO_2 from the atmosphere on a net basis; Biomass conversion systems, with or without the generation of heat, power and/or fuels, could be combined with a process which separates carbon. If the biomass is produced in a sustainable way and the separated carbon is subsequently stored away from the atmosphere for a very long period of time, the entire process becomes a continuous carbon sink, i.e. negative emissions are yielded. Net lifecycle emissions would be strongly negative because the carbon in the biomass was originally extracted from the atmosphere via photosynthesis and carbon will be extracted again when new biomass replaces what has been harvested. This feature distinguishes biomass from other carbon-lean energy sources, such as solar, wind, geothermal or nuclear energy. In this chapter these technologies are referred to as "negative emission biomass technologies". An other approach to yielding negative emissions would be air CO_2 capture, whereby ambient CO_2 is removed by passing a natural air flow over absorber surfaces [Lackner, 2003]. The CO_2 concentration in ambient air is around a factor of 100 or more lower than in flue gas. The IPCC [2005] conclude that capturing CO_2 from air by the growth of biomass and its use in industrial plants with CO_2 capture is more cost effective and based on foreseeable technologies. The capture of CO_2 from ambient air is not discussed further in this chapter.

This chapter deals with negative emission biomass technologies, which can be divided into two sub groups; centralised or distributed:

1. Centralised negative emission biomass technologies build on the conversion of biomass into CO_2-free or CO_2-lean energy carriers in centralised conversion plants. CO_2 is captured in the conversion plant and subsequently transported and stored in geological formations, see Fig.2. This is referred to as biomass energy with CO_2 capture and storage (BECS). For economic reasons the CO_2 capture, transportation and storage must be stationary and large-scale.

2. Distributed negative emission biomass technologies build on the conversion of biomass into long-term carbon-sequestering charcoal. The conversion may be combined with the production of biofuels. Carbon can be sequestered by utilizing the wood charcoal as a soil amendment in forests and arable lands to improve soil productivity.[3]

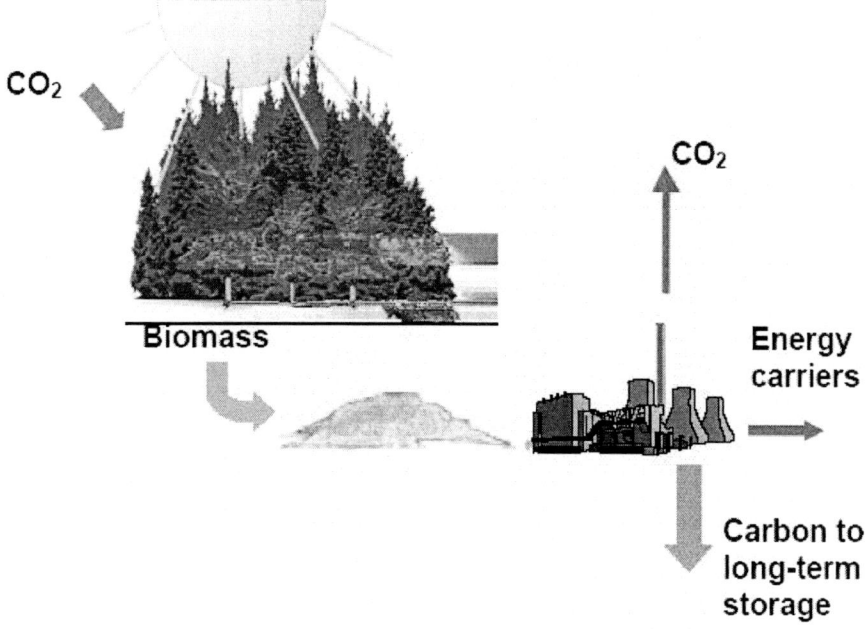

Figure 2. The principle of bioenergy with CO_2 capture and storage (BECS).

Early work on negative emission biomass technologies was presented by Seifritz [1993] and Williams [1998]. Seifritz [1993] suggested that massive implementation of charcoal sequestration in developing countries could be implemented to off-set fossil emissions from industrialised countries. It is argued in the paper that the strategy outlined would offer

---

[3] In addition, the option of "remote sequestration" exists, i.e. biomass may be harvested and separately sequestered, for example by burying the trees (see Keith, 2001, for further discussion concerning this alternative).

industrialised countries a relatively cheap way to pay the external costs of their GHG emissions. At the same time well-needed capital flows from industrialised to developing countries would be created. Williams [1998] analysed production of hydrogen from biomass combined with CCS. The author proposes that BECS could open up possibilities for achieving deep emission reductions globally even if certain countries are unable (or unwilling) to significantly reduce emissions. The solution, then, would be that negative emissions generated by one party could be used to permanently off-set emissions generated by another.

Obersteiner et al. [2001] brought a new dimension to the discussion surrounding negative emissions by arguing that removal of CO_2 from the atmosphere through massive deployment of BECS would make it possible for mankind to bring about a downwards movement of atmospheric CO_2 levels. The negative emission feature could thus be used to off-set emissions from the past. Following the observations by Obersteiner et al. [2001], modelling exercises simulating the global energy system have been carried out in order to assess the potential of BECS on a global scale [Azar et al., 2006; Smith et al., 2006, Rao and Riahi; van Vuuren et al.]. The importance of considering BECS when judging the feasibility of ambitious GHG stabilisation targets was illuminated in a study by Azar et al. [2006] which concluded that the option of CCS applied to fossil fuels and bioenergy could reduce the cost of meeting a 350 ppm stabilisation targets by 75 % compared to a case where these technologies are unavailable. According to the study, the corresponding number when CCS was applied exclusively to fossil fuels was 45 %.

A rapid downwards movement of atmospheric CO_2 levels could prove to be a necessity of a mitigation strategy in conformity with the United Nations Framework Convention on Climate Change (UNFCCC) [Obersteiner et al., 2004]. The ultimate objective of the UNFCCC [1992] is a "Stabilization of greenhouse gas concentrations in the atmosphere at a level that would prevent dangerous anthropogenic interference with the climate system". This level should be achieved within a time frame sufficient to, inter alia, allow ecosystems to adapt naturally to climate change. According to Thomas et al. [2004], gradual climate change can be expected to commit 24 % of terrestrial species to extinction by 2050 due to rising temperatures (on the basis of mid-range climate-warming scenarios). Thomas et al. [2004] argue that returning to near pre-industrial global temperatures as quickly as possible could prevent much of the projected climate-related (and irreversible) extinction from being realized. In other words, negative emission biomass technologies allowing for a downward movement in atmospheric CO_2 levels could be instrumental in achieving the ultimate objective of the UNFCCC.

Negative emission biomass technologies also have the potential to bring new perspectives to environmental product branding. Möllersten and Yan [2001] identified clearly negative CO_2 balances for BECS-based electricity production. Williams [1998] and Möllersten et al. [2003a] reported distinctly negative net life cycle CO_2 emissions for biofuels (ethanol, hydrogen and methanol). One could thus consider negative emission biomass technologies to be a technological possibility enabling the generation of energy carriers and other products that are "greener than green". Möllersten et al. [2006] and Day et al. [2005] reported clearly negative carbon balances with respect to the production of pulp/paper and hydrogen,

---

[4] Net power output/lower heating value of fuel input. The lower heating value (LHV) is used as the basis for the calculations and numbers presented throughout this paper.

respectively, when negative emission biomass technology is integrated in the production. Consequently, manufacturers could potentially co-generate specific products and atmospheric CO_2 removal, thus adding a new dimension to environmentally friendly products. Day et al. [2005] put forward the novel idea that when a product has a negative carbon budget on a net basis, consumerism "becomes an agent of climate mitigation".

Negative emission biomass technologies have a potential role to play as a component of a carbon management technology portfolio – a role that could not easily be replaced by any other technology. However, the assessment of the individual technologies, economic analysis has mostly been restricted to estimates concerning key numbers such as the cost of electricity (COE) or the cost of CO_2 captured. To our knowledge, no study has so far been carried out that assesses the commercial potential of an individual negative emission biomass technology taking into account the long-term development of energy and carbon markets. Furthermore, only few of many perceivable negative emission biomass technology system configurations (from biomass production to carbon storage) have been studied in any detail (for example, Williams, 1998; Möllersten et al., 2006; Okimori et al., 2003, Larson et al., 2006). One should note that negative emission biomass technologies form a technology cluster with a large number of perceivable system configurations. Only a few examples of these can be found scattered throughout the literature. This brings us to the aim and scope of this chapter;

Firstly, a BECS implementation scenario study is presented. The study analyses investments in BECS under multiple uncertainties in a pulp and paper mill environment. An introduction to pulp and paper production and related energy issues of interest to the study is followed by a section where the mill environments (market pulp mills and integrated pulp and paper mills) and the selected integrated BECS systems are defined. The results from an analysis of the energetic performance of the selected BECS systems are presented as well as an analysis of overall CO_2 balances. Uncertainty is considered in the economic modelling through the use of stochastically correlated price processes of one input price (biomass) and two output prices (electricity and CO_2 emission allowances) that are consistent with shadow price trajectories of a large-scale global energy model. The investment analysis is carried out within a real options framework (see, for example, Dixit and Pindyck, 1994). A major advantage of this approach is the ability to analyze the value of flexibility and uncertainty. The aim of this study is to present a comprehensive modelling effort targeted at evaluating the commercial potential of a negative emission biomass technology.

Secondly, having analysed the economic performance of BECS in a pulp and paper mill setting in some detail, we turn our focus to a broader discussion concerning negative emission biomass technologies. The aim is to identify important knowledge gaps and Research and Development challenges ahead of us.

EVALUATION OF CO_2 CAPTURE AND STORAGE IN PULP AND PAPER MILLS UNDER ECONOMIC AND TECHNICAL UNCERTAINTY

Combined Heat and Power Production in Pulp and Paper Mills

Pulp is used as a raw material to produce paper and board. Pulp production starts with a fibre source, the prime source being trees. Wood pulp is made by a mechanical or a chemical pulping process, or a combination of these two pulping processes called semi-chemical pulping. Paper and board production can be integrated with pulp production in integrated pulp and paper mills. In many cases, however, the pulp is produced in market pulp mills and then transported to another site where the paper or board is produced.

In chemical pulping wood chips are cooked in a solution of chemicals whereby the wood cellulose fibres, which are used for pulp production, are separated as the chemicals dissolve the lignin (delignification). A chemical pulping process called Kraft pulping accounts for around 70 % of pulp production worldwide [FAO, 2005]. The Kraft process generates a by-product from fibre extraction known as black liquor, which is a mixture of lignin and inorganic chemicals. The dissolved lignin in the black liquor represents just over half of the biomass entering a Kraft pulp mill.

In modern Kraft pulp mills the black liquor is burned in recovery boilers (RB), which recover important pulping chemicals and produce steam (by utilising the energy value of the dissolved biomass) that is fed to the mill combined heat and power (CHP) system. The efficient utilisation of the black liquor energy content can reduce the Kraft pulp and paper industry's reliance on fossil fuels. In the most energy efficient existing Kraft market pulp mills the fuel requirement for the CHP system, which satisfies the mill's requirement for medium-pressure (MP) and low-pressure (LP) steam, is typically covered through black liquor and internally generated bark. In contrast, integrated pulp and paper mills and paper mills need to import fuels to satisfy the process steam demand. In nearly all Kraft pulp production fossil fuels are still used in lime kilns, although a limited number of kilns were converted to biofuels in the 1980's [Siro, 1984]. Most pulp mills and all integrated mills rely on electricity import to cover the part of their electricity demand that is not covered by power generated with internal CHP.

In existing Kraft pulp mills with modern CHP systems based on recovery boilers and biomass boilers with steam turbines, electrical efficiencies are fairly low (up to 15 %[4]) [Larson et al., 2000]. Improved overall energy efficiency and increased electrical efficiency could be accomplished through the introduction of black liquor integrated gasification combined cycle (BLG/CC) [Berglin et al., 1999; Larson et al., 1999, Larson et al., 2000; Maunsbach et al., 2001], which is a promising technology that is however not commercially available today. Larson et al. [1999] modelled the performance of black liquor and biomass integrated gasification combined cycle in a typical present-day North American pulp mill. The results predict electrical efficiencies around 28–29 %.

Today, there is a trend in the pulp and paper industry towards a further "closing" of the process. This means minimising the amount of effluents together with reducing the need for additional raw materials and energy. Generally this will reduce the heat demand through improved heat integration. The higher power-to-heat ratio of BLG/CC compared to recovery boilers with steam turbines makes this development favourable for BLG/CC. Several studies have shown that CHP systems based on black liquor gasification (BLG) in mills with low energy requirements could turn the mills into net exporters of electricity (see, for example, Berglin et al., 1999; STFI, 2003).

Technologies for Reducing CO_2 Emissions in Chemical Pulp and Paper Mills

The awareness of carbon management is growing as the issue of anthropogenically induced climate change is gaining importance. Industry is looking for strategies, policies, and measures that could be adopted to address and reduce its GHG emissions and the pulp and paper industry is no exception, see, for example, Browne [2003], Bruce [2000] and Miner and Lucier [2004]. A large share of CO_2 emissions that can be allocated to pulp and paper production are closely related to the energy utilisation in pulp and paper mills. Opportunities for CO_2 reductions in Kraft pulp and paper mills can be organised in the following categories:

- Decreased specific energy utilization
- Fuel switching (to less carbon-intensive fossil fuels or biomass fuels)
- CO_2 capture and storage (CCS)

Possible projects range from highly process-integrated, as in the case of energy conservation through the optimisation of black liquor evaporation to pure energy projects, such as switching from fossil fuels to bark for steam generation (see, for example, Martin et al., 2000; Upton and Mannisto, 2001; Möllersten, 2002). When fossil fuels have been phased out, the main remaining mitigation option is BECS. BECS would lead to avoided on-site emissions thus potentially generating eligible tradable emission permits for the mill owner.[5]

Several studies analysing the potential for implementing BECS in pulp and paper mills have been published [Ekström, 1997; Möllersten and Yan, 2001; Möllersten, 2002; Möllersten et al., 2003a-b; Möllersten et al., 2006]. Before the results from these studies are summarised a brief background on CCS is provided.

A Background on CO_2 Capture and Storage (CCS)

The principle of CCS is to prevent CO_2 from escaping to the atmosphere by implementing technologies that separate, or "capture", the CO_2 from fuel conversion and store CO_2 or carbon in some form for long periods of time. CO_2 capture, transportation and storage technologies are feasible and technically proven. There is considerable experience accumulated in the chemical and petroleum industries for operating chemical reactors and absorption units used for the capture of CO_2 as well as for CO_2 transportation systems. CO_2 is routinely separated today at some large industrial plants such as natural gas processing and ammonia production facilities, although these plants separate CO_2 to meet process demands and not for storage [IPCC, 2005]. In its third assessment report, the IPCC [2001b] predicts that CCS technologies could give major contributions to CO_2 abatement by 2020 provided that the integrity of storage can be guaranteed.

CO_2 Capture

CO_2 capture applied to emissions from biomass can be divided into process groups with respect to the capture technology that is used. The purpose of CO_2 capture is to produce a concentrated stream which can readily be transported to a CO_2 storage location. Here, we choose to focus the description on three main process groups, see Fig. 3.

---

[5] An accounting system must be in place that enables crediting of avoided emissions through BECS (see section XX for further discussion about issues related to the accounting of captured and stored biotic CO_2).

"Post-combustion capture", the capture the CO_2 from combustion products in flue gas, is the most conventional approach for CO_2 capture. CO_2 capture from process streams is an established concept that has achieved widespread industrial application. These applications have, however, focused on gas separation from high purity, high pressure streams where the energy penalties and cost for capture are moderately low. Post-combustion capture for CO_2 abatement involves capturing CO_2 from gas with low CO_2 concentration. This means that a large amount of inert gas has to be treated which leads to a significant cost and efficiency penalty because of the size of any downstream scrubbing and heat recovery equipment, etc. A main challenge associated with post-combustion capture for abatement of GHG emissions is reducing costs and the amount of energy required for capture.

"Pre-combustion capture" systems share a common objective: to produce a fuel stream that contains little, or none, of the carbon contained in a carbonaceous feedstock fuel. This approach necessarily involves the separation of CO_2 at some point in the conversion process. The resulting H_2-rich fuel can be fed to a hydrogen consuming process such as production of synthetic liquid fuels, oxidised in a fuel cell, or burned in the combustion chamber of a gas turbine to produce electricity. For solid fuels like biomass (or coal), the first step in a pre-combustion system is always a gasification process, by which the solid fuel is reacted with steam and/or oxygen to produce a fuel gas that contains large quantities of carbon monoxide (CO) and hydrogen (H_2). The synthesis gas is cleaned and the CO would be reacted with steam ("CO/water-shift reaction") to produce more H_2 and CO_2, thereby increasing the amount of CO_2 available for capture.[6] The separation of these two gases can be achieved with well known, commercial absorption-desorption methods, producing the CO_2 stream suitable for storage. These technologies have a high strategic importance because their capability to deliver a suitable mix of electricity, hydrogen and lower carbon-containing fuels. Conversion of carbonaceous fuels to synthesis gas, CO/water-shift conversion and CO_2 separation are well-known processes which could be applied in power stations.

"Oxygen combustion" (often referred to as "oxy-fuel") is an approach that builds on the combustion of a fuel using pure oxygen instead of air. The dilution of CO_2 with nitrogen in the flue gases is thus avoided. Consequently, the flue gases will consist of CO_2 and steam, which in turn enables the CO_2 to be separated by condensing the steam. The oxygen combustion approach will require a considerable amount of energy for the production of pure oxygen.

CO_2 Transportation

Because of the large volumes involved in CCS, pipelines are suitable for the transportation of CO_2 to a storage location once it has been captured. Transport of CO_2 can best be done at high pressure in the range of 80 to 140 bars. Compression and pipeline transport of CO_2 is feasible and technically proven. Several million tons of CO_2 are transported annually, mainly in the USA, over long distances on-shore in pipelines for use in the enhanced oil recovery industry [IEA, 2002]. Economies-of-scale effects are significant on the cost of CO_2 transportation by pipeline. This is illustrated in Fig. 4 which illustrates the costs of

---

[6] $CO + H_2O_{vap} \rightarrow CO_2 + H_2 +44.5$ MJ/Mol$_{co}$

transportation and on-shore storage as a function of distance for a set of CO_2 flow rates.[7] The cost data presented in the diagram were generated using two independent cost models [IEA, 2002; Ogden, 2002]. This is a major reason why CCS is only feasible at larger scales.

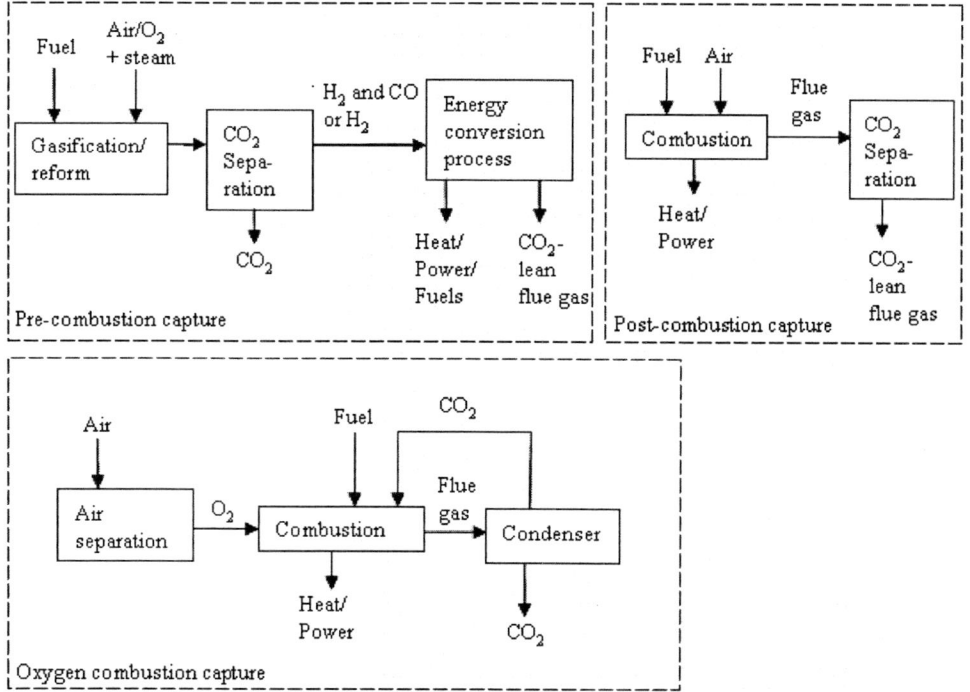

Figure. 3. Schematic representation of capture systems.

Other means of transportation that can be used are motor carriers, railway and water carriers.

Experiences from these means of transportation are mainly found in the food and brewery industry, and the amounts transported are in the range of some 100,000 tons of CO_2 annually, which is much smaller than the amounts associated with CCS [Svensson et al., 2004]. Svensson et al. [2004] conclude that pipeline and water carriers and a combination of these are the only economically feasible alternatives.

CO_2 Storage

A key issue is where CO_2 should be stored. The discussion on CO_2 storage covers the injection of supercritical-state CO_2 into underground geological formations or the deep oceans and technologies for converstion to stable carbonates or bicarbonates. Much further work is required to investigate the permanent storage of CO_2. Deep underground storage is regarded as the most mature storage option today [IPCC, 2005]. Candidate underground storage locations are exhausted natural gas and oil fields, not exhausted oil fields, unminable coal formations, and deep, saline water bearing formations, see Fig. 5. Several commercial

---

[7] CO_2 injection is assumed to take place in CO_2-retaining deep saline aquifers and the depth of the injection wells is 1000 m. Capital costs were annualised using an interest rate of 10 % and a plant life of 25 years. A capacity utilization of 90 % was assumed.

projects involving the injection of CO_2 into reservoirs where it displaces and mobilises oil (so-called enhanced oil recovery) are in commercial operation [IPCC, 2005]. Underground storage is characterised by minimum interference with other ecological systems and provision of storage for very long time periods whereas ocean storage has considerable uncertainties regarding potential environmental damage, especially effects on marine life due to increased acidity, and the long-term isolation of the CO_2 [Falkowski et al., 2000; IPCC, 2005]. International monitoring of current storage projects will help to define criteria and standards for safe geological CO_2 storage.

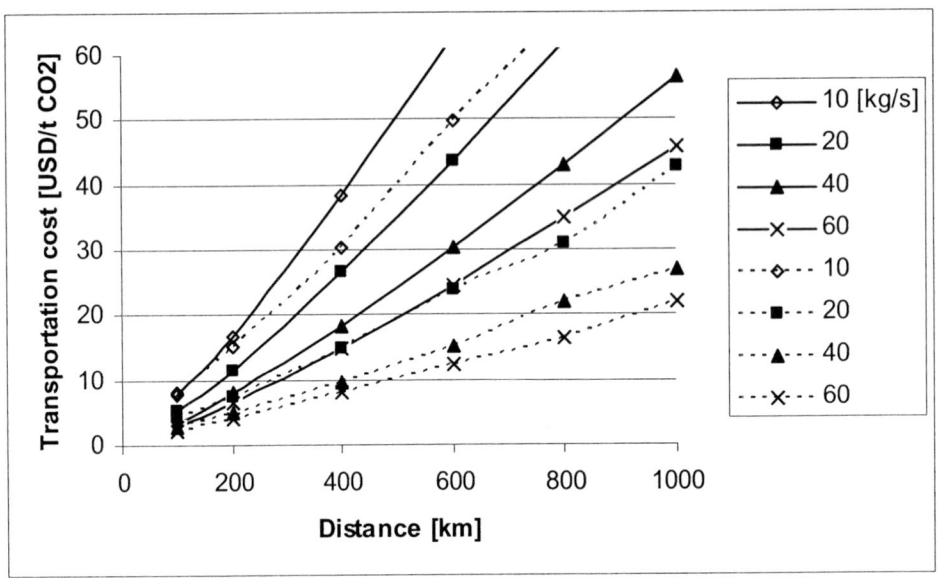

Figure 4. CO_2 pipeline transportation costs according to two independent cost models by IEA [2002] (dotted line) and Ogden [2002] (full line).

CO_2 could also be reacted with certain minerals and subsequently be stored as carbonates (stable products that are common in nature) [Lackner, 2003; DOE, 1999]. The mineralization option is, however, flawed by high costs and energy requirements with current best approaches [Baciocchi et al., 2006]. Improved methods for accelerating carbonation are needed.

One major concern with underground storage is the possibility of leakage of the stored CO_2. Leakage rates have to be very small for carbon capture and storage to play a large and meaningful role in global efforts to meet stringent climate targets.[8] For more details on the leakage risk, see, for example, Ha-Duong and Keith [2003].

---

[8] If CCS would become a major option for the abatement of CO_2 emissions very large quantities of CO_2 would be stored towards the end of this century. If, for example, 1 % of 600 GtC (in line with the quantity stored globally the year 2100 according to some modelling exercises) would leak from storage every year, total emissions from leakage would amount to as much as 6 GtC/year, which is roughly equal to current total global CO_2 emissions from fossil fuels.

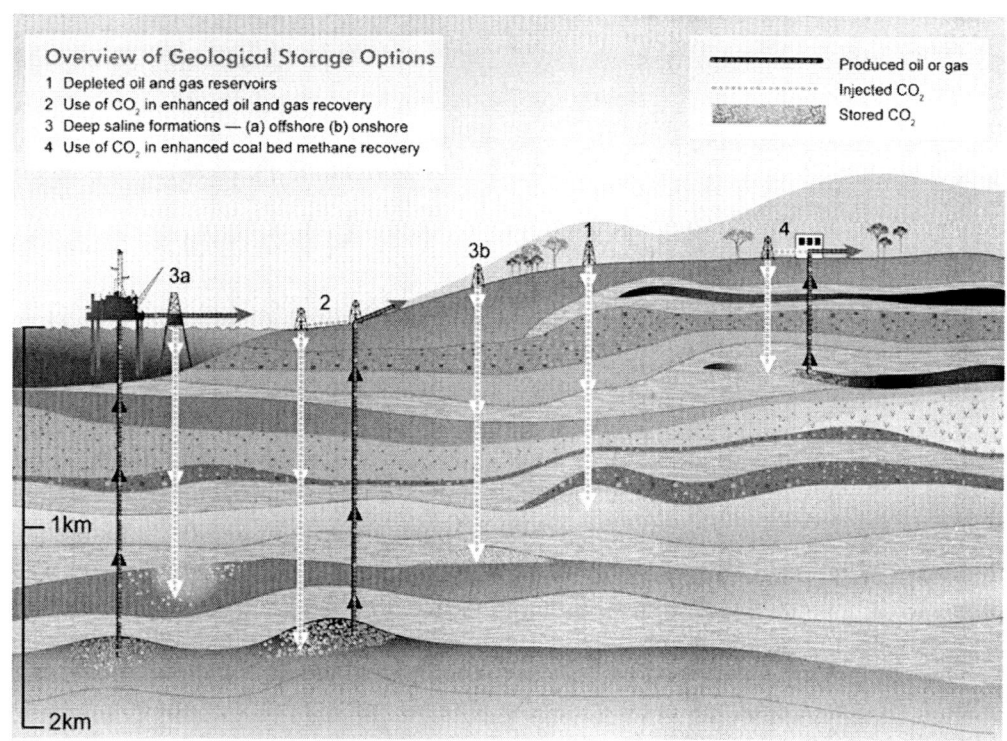

Figure 5. Overview of geological storage options (Source: IPCC, 2005).

Several estimates of the global underground storage potential have been carried out. In Table 1, the comprehensive assessment by the IPCC [2005] is reproduced.

Table 1. Potential for geological carbon storage options

Reservoir type	Lower estimate of storage capacity (GtCO$_2$)	Upper estimate of storage Capacity (GtCO$_2$)
Deep saline formations	1000	Uncertain, but possibly 10 000
Oil and gas fields	675	900
Unminable coal seams	3 - 15	200

Source: IPCC [2005]

CO_2 could also be reacted with certain minerals and subsequently be stored as carbonates (stable products that are common in nature) [Lackner, 2003; DOE, 1999]. The mineralization option is, however, flawed by high costs and energy requirements with current best approaches [Baciocchi et al., 2006]. Improved methods for accelerating carbonation are needed.

One major concern with underground storage is the possibility of leakage of the stored CO_2. Leakage rates have to be very small for carbon capture and storage to play a large and meaningful role in global efforts to meet stringent climate targets.[9] For more details on the leakage risk, see, for example, Ha-Duong and Keith [2003].

CO_2 Capture and Storage in Pulp and Paper Mills

The performance of BECS in pulp and paper mills has been the focus of some published works. Here, we first provide a summary of studies based on CHP systems of existing-standard market Kraft pulp mills followed by results concerning future-standard mills. Ekström et al. [1997] analysed post-combustion capture and oxygen combustion in pulp mill CHP systems based on recovery boilers and steam turbine technology. The investment cost required for the entire CHP system was estimated to double due to the introduction of CO_2 capture. Möllersten and Yan [2001] analysed advanced BLG-based systems with CCS for co-production of biomass-based transportation fuels, power, and heat. Clearly negative CO_2 balances were identified. Furthermore, the systems' rates of return on investment (ROR) were compared in order to determine the price of CO_2 emission permits that would justify the extra costs required for the more advanced and costly systems compared to a reference case based on recovery boiler and steam turbine technology (without CCS). The analysis considered potential incomes for both energy products delivered and from trading of CO_2 emission permits. It was found that, assuming a 700 km CO_2 transportation requirement from the pulp mill to the injection site, a price of approximately 60 \$/tCO_2 justified the introduction of the advanced systems with CCS.[10] Möllersten et al. [2003] evaluated CO_2 abatement potentials of CCS in CHP systems based on recovery boilers and pressurised BLG/CC, respectively. The largest abatement potential found was for post-combustion CO_2 capture by chemical absorption from recovery boiler and bark boiler flue gases. For the analysed BLG/CC systems with pre-combustion capture, the analysis was restricted to partial CO_2 capture without a CO/water-shift reaction prior to the CO_2 separation by chemical absorption. According to the study BLG/CC with partial pre-combustion CO_2 capture features a higher electrical efficiency compared to boilers with post-combustion capture but a significantly lower overall CO_2 abatement potential. One BECS study has been published that takes into consideration the potential of energy efficiency improvements in pulp and paper mills. Möllersten et al. [2006] investigated the integration of CHP systems with CCS in predicted future market pulp mill and integrated pulp and paper mill environments ("reference mills") with considerably lower process steam demand than existing-standard mills. The studies considered three main types of CHP systems with CO_2 capture: post-combustion capture from recovery boiler flue gases, pre-combustion capture in BLG/CC systems without CO/water-shift enhancement, and pre-combustion in BLG/CC systems with CO/water-shift enhancement. A few conclusions can be drawn from a comparison of the analysed systems; Considerably less CO_2 is captured in BLG/CC systems without a CO/water-shift reaction than in the other two systems considered in the study. Meanwhile, the CO_2 capture level in BLG/CC systems with CO-water shift reaction and post-combustion capture from recovery boiler flue gases were approximately the

---

[9] If CCS would become a major option for the abatement of CO_2 emissions very large quantities of CO_2 would be stored towards the end of this century. If, for example, 1 % of 600 GtC (in line with the quantity stored globally the year 2100 according to some modelling exercises) would leak from storage every year, total emissions from leakage would amount to as much as 6 GtC/year, which is roughly equal to current total global CO_2 emissions from fossil fuels.

[10] Note that the method used for the evaluation is not consistent with the design of CO_2 accounting principles in emerging emission trading systems, e.g. the EU Greenhouse Gas Emission Trading Scheme (EU-ETS).

same. The studies further showed that CO_2 capture systems based on pre-combustion capture have advantages compared to post-combustion capture systems in terms of higher electrical efficiency and lower biomass fuel consumption.

Furthermore, estimated costs of CO_2 capture can be reported from the studies mentioned above. Möllersten et al. [2003a] estimated the cost of CO_2 capture to 34 USD/tCO2 for the case of post-combustion capture. For pre-combustion capture the costs in the range 22 and 34 USD/tCO$_2$ were estimated [Möllersten et al, 2003; Möllersten et al., 2006].

Economic Evaluation Under Uncertainty

Investment strategies in capital- and energy-intensive industries, like the pulp and paper industry, are driven by long-run price signals and their respective uncertainties. The implementation of climate policies is a major source of uncertainty for these industries. In the economic evaluation presented here, BLG-based[11] CHP systems with a CCS option in predicted future-standard Kraft pulp and paper mills are evaluated against market conditions predicted by large scale global energy models.[12] The economic feasibility of BECS is evaluated given correlated uncertainties of the biomass fuel, electricity and CO_2 emission permit prices.[13] Uncertainties in connection with climate change, and society's response to the threat of climate change, along with the irreversible characteristics of many investments in mitigation technologies, create the conditions for decision-makers to value delaying investment decisions until more information is available. For the valuation of the investment decision we will build on principles of the real option theory (see, for example, Dixit and Pindyck, 1994; Mun, 2002).

In the next section the evaluated CHP systems are defined. Based on simple process simulations, the performance of the evaluated systems are then presented, emphasising the rates of electricity production, CO_2 capture and biomass consumption. Some effort is made to estimate and discuss overall CO_2 balances of the evaluated systems, taking into account primary ("on-site") and secondary ("off-site") impacts on the emissions.[14] The capital costs of the evaluated CHP systems are estimated in the following section. Next, the modelling

---

[11] This case study does not address the issue whether recovery boilers or BLG can be expected to be the preferred technology in the future, but simply assumes the BLG as the baseline technology. As mentioned earlier, the currently applied process for recovery of chemicals and energy from black liquor is based on recovery boilers with Rankine steam cycle. If and when BLG will be introduced in reality is dependent on a number of factors. The recovery boiler with Rankine steam cycle has some drawbacks; the thermal efficiency and power-to-heat ratio are low, capital cost is high, maintenance under corrosive conditions is complicated, and there is a risk of smelt/water explosions. BLG is addressing the majority of these drawbacks. In addition, BLG opens up the opportunities for tailoring the delignification process towards higher yields and/or improved pulp physical properties [Stigsson and Berglin, 1999]. The gasification technology development has taken place during the last two decades. A number of BLG technologies have been developed, e.g., the Chemrec, MTCI, and ABB processes. In an on-going 10 M€ project the Chemrec BLG technology is demonstrated in Sweden.

[12] Clearly, the results have to be interpreted with consideration to implicit assumptions made thereby (the conditions underlying the scenario family chosen etc.).

[13] Consider that it is probable that more stringent CO_2 restrictions will not only lead to an increase in the CO_2 price, but also lead to elevated electricity and biomass prices. One realises that as the rate of CO_2 capture, electricity production, and biomass fuel consumption are correlated, this has implications for the economic feasibility of the technologies and their relative competitiveness.

[14] However, it is only the direct impacts that matter to the economic evaluation, as the mill owner is only accounts for the "own" emissions within the framework of an emission trading system.

framework, that optimises the pulp mill or integrated pulp and paper mill owner's decisions, is described. Finally, the results of the economic evaluation are presented.

Definition of the Evaluated CHP Systems

In this section, the mill environments and CHP system configurations evaluated in this study are defined. The systems' energy performance and overall CO_2 balances are presented.

Mill Environment

The modelling of CHP systems in this study is carried out in two different mill environments: a market pulp mill (MPM) and an integrated pulp and paper mill (IPPM). The MPM is identical with the model MPM defined by the Swedish Ecocyclic Pulp Mill research programme [STFI, 2003]. In the MPM of the Ecocyclic Pulp Mill Programme, which is assumed to employ the best technologies available in commercial use in the late 1990's in all departments of the mill, the required process steam is 11 GJ/ADt pulp (Air-Dry tonne pulp) ⚸ a reduction by 24 % compared to the 1994 Swedish average. The MPM has a capacity to produce 1550 ADt pulp/d.

The IPPM, defined by Berglin et al. [1999], is an extension of the Ecocyclic Pulp Mill programme MPM. The IPMM steam consumption is approximately 5 % lower than the average Swedish 1994 fine paper mill. The IPPM has a capacity to produce 1860 tonnes of paper/d.

The process steam and electricity requirements of the MPM and IPPM are presented in Table 2.

Table 2. Energy requirements of the considered mills

Energy requirement (GJ/ADt end product)		
	Market pulp mill	Integrated pulp and paper milla
Electricity	2.5	4.8
Medium pressure steam (12 bar)	4.3	7.5
Low pressure steam (4 bar)	5.7	8.3

[a] 1.2 tonnes paper are produced for every air-dry tonne (ADt) pulp produced in the IPPM.

CHP System Configuration

Table 3 summarizes the alternative CHP system configurations considered. All cases are based on a pressurised high-temperature, oxygen-blown black liquor gasifier. In all evaluated systems the synthesis gas is cooled in a quenching bath using the weak wash as coolant whereby the weak wash is evaporated using the sensible heat of the synthesis gas. The quenching adjusts the fraction of steam in the synthesis gas to ensure an adequate amount of water for a CO/water-shift reaction to proceed in a downstream CO/water-shift reactor. Prior to shift-conversion, further gas cleaning would be required in order to protect the shift reactor from contamination. Gas cleaning was not modelled in this study.

Table 3. Summary of analysed CHP system configurations

	Black liquor conversion	Biomass conversion[a]	CO2 caputre[b]		
Case	Gasifier	Gasifier	None	No CO/water-shift	CO/water-shift
MPM/BLG$_1$	X	X	X		
MPM/BLG$_2$	X	X		X	
MPM/BLG$_3$	X	X			X
IPPM/BLG$_1$	X	X	X		
IPPM/BLG$_2$	X	X		X	
IPPM/BLG$_3$	X	X			X

[a] Defines the technology used when fuel in addition to black liquor is required to meet process steam demands.

[b] CO_2 capture from both black liquor and bark/woody biomass is considered when applicable.

In the MPM/BLG$_3$ and IPPM/BLG$_3$ cases the cooled synthesis gas is sent to a shift unit to adjust the CO/H$_2$ ratio via a CO/water-shift reaction. The shift unit is divided into two stages in series: The first (high-temperature) stage the temperature ranges from 225°C to 470°C. Most of the shift reaction is accomplished in this reactor. After being cooled down to 225°C, the synthesis gas is sent into the second (low-temperature) stage. The heat released in the shift unit is recovered through generating MP steam, which can be made useful for pulp and paper production. Sensible heat (above 70°C) of the synthesis gas leaving the shift unit is used to generate LP steam and some feed water for the heat recovery steam generator (HRSG). Subsequently the synthesis gas is cooled to 33°C before entering the cleanup unit.

In the MPM/BLG$_2$, MPM/BLG$_3$, IPPM/BLG$_2$, and IPPM/BLG$_3$ cases, CO_2 separation is carried out in physical absorption units (meant to approximate the Selexol process) upstream from the gas turbine combustion chamber. The captured CO_2 is subsequently compressed to 80 bar in a multi-stage intercooled compressor.

After the clean-up section the synthesis gas is used to fuel a gas turbine for power generation.[15] The exhaust gas from the gas turbine is recovered in a HRSG and the generated steam is used for process steam needs in the mills, either directly or via a back-pressure steam turbine which generates additional electricity.

When additional fuel is required to satisfy the process steam demand a supplemental BIG/CC is considered (see Table 3). In the same way as stated above, the biomass gasifier is modelled as a "black box" The main assumptions of the CHP systems are given in Table 4. Table 5 presents the main characteristics of the CO_2-lean fuel gas which is fed to the gas turbine after the physical CO_2 absorption.

---

[15] There are a few areas relating to the use of gas turbines in the studied systems where technical development is less advanced. BLG/CC without CO_2 capture requires that burners be modified for low and medium heating value gases. Fortunately, BLG/CC can draw on burner development for other gasification applications (e.g. coal and biomass) [Stigsson and Berglin, 1999]. The use of decarbonised fuels in gas turbine systems presents new technical challenges. Further development of gas turbines that can operate at very high inlet temperatures will be necessary to enable the firing with hydrogen-rich gas.

Table 4. Main assumptions for the evaluated CHP systems

Gasifiers				
	Black liquor		Biomass	
Cold gas efficiency (%)	77		77	
Synthesis gas properties				
	Raw gas	After quench	Raw gas	After quench
Temperature (°C)	950	211	900	209
Pressure (bar)	32	25	27	25
Composition (mol%)				
N_2	0.2	0.1	0.2	0.1
CO	29.5	13.5	30.0	13.0
CO_2	14.6	6.7	24.2	10.4
H_2O	22.0	64.3	15.9	63.7
H_2	31.1	14.2	24.1	10.4
H_2S	1.5	0.7	0.0	0.0
CH_4	1.1	0.5	5.6	2.4
Gas turbine[a]				
Turbine inlet temperature (°C)		1250		
Pressure ratio		17		
Mechanical efficiency (%)		98		
Isentropic efficiency, expander (%)		92		
Isentropic efficiency, compressor (%)		87		
Steam cycle				
Turbine inlet temperature (°C)		440		
Turbine inlet pressure (bar)		66		
Mechanical efficiency (%)		98		
Isentropic efficiency, expander (%) (High pressure / Medium pressure)		85 / 87		
Pinch temperature difference of HRSG (°C)		15		
Feed water temperature (°C)		120		

[a] Commercial gas turbines come in well-defined sizes that cannot be changed. Here, however, we refer to a "generic" gas turbine with the characteristics in Tab. 3.

Table 5. Characteristics of CO_2-lean fuel gas to the gas turbine[a]

Temperature (C)	110
Pressure (bar)	20
Composition (mol%)	
N_2	0.4
CO	0.4
CO_2	0.7
H_2O	0.0
H_2	96.8
H_2S	0.0
CH_4	1.8

[a] The composition is based on MPM/BLG2 and would vary slightly when gasified woody biomass is added as in IPPM3.

CHP System Performance

Simple simulations of CHP systems were carried out by using the ASPEN PLUS process simulator [AspenTech, 2003]. The entire energy system modelling and optimisation as well as the cost analysis could be done in much greater detail. We regarded, however, the detail of the energy system assessment in this case study to be sufficient for our level of systems analysis.[16] The performance of the analysed CHP systems is summarised in Tables 6 (MPM cases) and 7 (IPPM cases).

Table 6. Performance of the MPM CHP systems (Pulp production 1550 ADt/d)

	MPM/BLG$_1$	MPM/BLG$_2$	MPM/BLG$_3$
Black liquor (MW)	338		
Bark and woody biomass (MW)	0	0	0
CO$_2$ recovery (%)	0	31	90
CO$_2$ capture rate (kg CO2/s)	0	10	27
MP steam to mill (12 bar-t/h)	101		
LP steam to mill (4.5 bar-t/h)	137		
Power consumption for CO$_2$ absorption (MW)	N.A.	2	4
Heat consumption for CO$_2$ separation (MW)	N.A.	N.A	N.A.
Internal power consumption			
CO$_2$ compressor (MW)	N.A.	4	13
Air separation unit (ASU) (MW)	5	5	5
Others (MW)	10	10	10
GT output (MW)	100	99	93
ST output (MW)	21	16	10
Net electricity output (MW)	106	94	71
Mill electricity consumption (MW)	39		
Electricity surplus (MW)	67	55	32
Electricity surplus (MWh/ADt pulp)	1.0	1.0	0.5
Electrical efficiency (%)	31	28	21
Total efficiency (%)	76	72	65

The tables show the mill-integrated systems' performance with regard to fuel requirement, CO$_2$ capture rate (when applicable), electricity production and overall energy efficiency. Note that in all cases the mills' process steam demand is satisfied precisely. All MPM CHP systems generate a net electricity surplus which allows for power export to the

---

[16] The gasification itself is not simulated since the purpose of this study is to investigate the energy system not the gasification. The black liquor gasifier is treated as a "black box" in the simulation. However, the energy and material balances in the gasification have been considered in the simulation and the composition of the synthesis gas generated is correct. Only the parts that interact with the rest of the system are included such as compressors, heat exchangers and steam generators where process steam is produced. A large fraction of the sulphur in the black liquor is converted to hydrogen sufide (H2S) and is recovered and turned into useful sulphur for economic reasons. The H2S is removed from the gas in an acid gas removal system downstream from the gasifier. Data on the gasifier were taken from Berglin et al. [1999]. The composition of the raw synthesis gas from the gasifier was obtained from Lindblom [2001]. The high-temperature stage of the shift reactor was modelled as adiabatic reactor and the second stage as a constant temperature reactor. In the same way as stated above, the capture unit is modelled as a "black box". The work consumed for the physical absorption depends on the partial pressure of the CO2 in the gas mixture. The work required for operating the absorption plant was set to 0.14 MJ/kgCO2 captured.

grid, although the electricity surplus drops when CO_2 is captured. Moreover, we can see that process steam requirements are satisfied through the black liquor-based CHP system in the MPM cases. Also in the IPPM cases an electricity surplus is generated for all cases. In contrast to the MPM cases, however, the process steam requirements are not satisfied by the black liquor alone in any of the IPPM cases.[17]

Table 7. Performance of the IPPM CHP systems (Paper production 1860 ADt/d)[a]

	IPPM/BLG$_1$	IPPM/BLG$_2$	IPPM/BLG$_3$
Black liquor (MW)	338		
Bark and woody biomass (MW)	114	114	184
CO_2 recovery (%)	0	33	90
CO_2 capture rate (kg CO2/s)	0	14	45
MP steam to mill (12 bar-t/h)	176		
LP steam to mill (4.5 bar-t/h)	200		
Power consumption for CO_2 absorption (MW)	N.A.	3	6
Heat consumption for CO_2 absorption (MW)	N.A.	N.A.	N.A.
Internal power consumption			
CO_2 compressor (MW)	N.A.	6	20
Air separation unit (ASU) (MW)	6	6	7
Others (MW)	14	14	16
GT output (MW)	135	135	146
ST output (MW)	0	0	16
Net electricity output (MW)	115	107	113
Mill electricity consumption (MW)	74		
Electricity surplus (MW)	42	33	39
Electricity surplus (MWh/ADt paper)	0.5	0.5	0.5
Electrical efficiency (%)	25	24	22
Total efficiency (%)	78	76	68

The results obtained from our simulation of BLG/CC are consistent with other studies evaluating BLG/CC (without CO_2 capture) in the same mill environment. For example, Berglin *et al.* (1999) reported electrical efficiencies in the range 29-31 % and 25-27 % for market pulp mills and integrated pulp and paper mills, respectively. We are not aware of published studies analysing CCS in similar mill environments that could be used for the sake of comparison.

---

[17] We can also observe that CCS leads to a larger drop in electrical efficiency for the MPM than for the IPPM. This can be regarded as an effect of the system configurations chosen and that the systems were optimised with respect to the mills' steam requirements rather than with respect to the total electrical efficiency. Note that an optimisation of a total energy system with the objective to maximise the electrical efficiency may lead to different results.

Impact of CCS on Overall CO_2 Emissions

It is only the direct emissions from a mill, and the impact that CCS would have on these, that would be relevant in relation to an emission trading system. However, the overall impact on the emissions is relevant with respect to the environmental credibility of a mitigation option.[18] In order to calculate the overall CO_2 reduction impact of introducing CCS in pulp and paper mills we consider changes in on-site and off-site net emissions compared to the reference cases MPM1 and IPPM1 following the principles outlined below.[19] The MPM/BLG$_3$ and IPPM/BLG$_3$ cases were selected for this analysis. The systems' overall CO_2 impact compared to the reference cases was assessed based on the performance of the systems presented in Tables 6 and 7.

The only change in on-site emissions compared to the reference cases is the CO_2 which is captured and put into long-term storage and thereby not allowed to reach the atmosphere.

Regarding off-site emissions the following approach was applied:

- The mills' electricity production from CHP is affected by introduction of CO_2 capture. Reduced or increased electricity production will have an impact on the electricity balance between the mill and the grid. We assume that the marginal electricity supply which has to compensate for a change in the electricity balance comes from either natural gas-fired combined cycle (NGCC) power plants with a 60 % electrical efficiency[20] or coal-fired condensing power plants with 40 % electrical efficiency.

- Extraction of biomass requires energy, which leads to net emissions of CO_2 if fossil fuels are used. The internally generated black liquor available at the mills is a by-product of the pulp production which has no alternative use. Consequently, the emissions due to the extraction of the black liquor biomass fraction should be allocated to the production of pulp and paper. Therefore, only emissions from the extraction of biomass required in excess of the available black liquor are allocated to the CHP systems in our analysis. It is reasonable to consider two alternative levels of CO_2 emissions for biomass extraction. As a lower value, we considered data for unrefined forestry residues. Börjesson and Gustavsson [1996] estimate that 2.9 kg CO_2 is emitted per GJ forestry residues extracted. The figure, based on Swedish conditions, includes 50 km transportation of the fuel. As a value on the high end we used data for dedicated biomass plantations. The result of a comprehensive environmental life cycle assessment of fuel supply from dedicated eucalyptus plantations shows that 21 kg CO_2 is emitted per GJ biomass extracted [Dowaki et al., 2002].

- CO_2 transportation by pipeline requires work for pressurisation. The initial pressurisation is considered in our analysis in that compression penalises the net power output of the mill CHP systems. The impact of emissions due to work required for booster compressors along the pipelines is regarded as negligible.

---

[18] Doubts are often expressed whether large-scale biomass systems can really mitigate climate change if life cycle emissions are taken into account, such as CO_2 emitted during biomass cultivation, harvesting, transportation and processing. Similar doubts have been expressed concerning BECS.

[19] Note that all CHP systems are analysed in an environment of predicted future pulp and paper mills with considerably lower process steam demand than today's existing mills. Thus the results presented do not reflect the advantages of the predicted pulp and paper mills that are per se energetically superior to existing mills.

[20] Representing the best available technology for natural gas-fired combined cycle power plants today and in the near future.

- CCS storage requires additional infrastructure such as pipelines. It is important to ensure that emissions are not moved from the tailpipe to the construction process to any significant extent. Comprehensive life cycle assessments of large-scale hydrogen production with CCS show only negligible CO_2 emissions due to the construction of additional infrastructure [Strømman, 2003]. The impact of these emissions was regarded as negligible in this study.

The resulting overall CO_2 impact of introducing CCS based on these assumptions is presented in Tables 8 and 9 (marginal electricity from NGCC and coal-fired power plants, respectively) as tonnes CO_2 saved per air-dry tonne of final product (pulp or paper). In Tables 8 and 9 one can also see the relative importance of the different emission sources. The results show that the CO_2 penalty, which derives from off-site emissions is quite small.

Table 8. Impact on CO_2 emissions compared to reference case (marginal electricity from NGCC)

Case	On-site emissions compared to reference (tCO_2/ADt)	Off-site emissions compared to reference (tCO_2/ADt)		Overall emissions compared to reference (tCO_2/ADt)[a]	Limited biofuel model (tCO_2/ADt)
	CO_2 capture and storage	Marginal electricity production	Biomass fuel extraction[a]		
MPM2[b]	-1.51	0.18	0[d]	-1.37/-1.37	-1.37[d]
IPPM3[c]	-2.15	0.02	0.01/0.08	-2.12/-2.05	-1.75

[a] First value based on higher biomass extraction CO_2 emission level/Second value based on lower biomass extraction CO_2 emission level

[b] MPM1 is reference case. CO_2 emissions are per ton pulp.

[c] IPPM1 is reference case. CO_2 emissions are per ton paper.

[d] No additional biomass fuel is extracted

Table 9. Impact on CO_2 emissions compared to reference case (marginal electricity from coal-fired power plants)

Case	On-site emissions compared to reference (tCO_2/ADt)	Off-site emissions compared to reference (tCO_2/ADt)		CO_2 emissions compared to reference (tCO_2/ADt)[a]	Limited biofuel model (tCO_2/ADt)
	CO_2 capture and storage	Marginal electricity production	Biomass fuel extraction[a]		
MPM2[b]	-1.51	0.45	0[d]	-1.06/-1.06	-1.06[d]
IPPM3[c]	-2.15	-0.05	0.01/0.08	-2.19/-2.12	-1.82

[a] First value based on higher biomass extraction CO_2 emission level/Second value based on lower biomass extraction CO_2 emission level.

[b] MPM1 is reference case. CO_2 emissions are per ton pulp.

[c] IPPM1 is reference case. CO_2 emissions are per ton paper.

[d] No additional biomass fuel is extracted

So far in this analysis the net emissions of CO_2 from biomass fuel without CO_2 capture have been considered to be zero. The real net emissions depend upon the carbon storage in soil and trees, the time span considered, the alternative use of the biofuel, and other factors [Schlamadinger et al., 1997]. Karlsson [2003] and Grönkvist et al. [2003] argue that the "wood-fuel efficiency" should be included into the analysis of biomass energy utilisation. If biofuel becomes a scarce resource, the limited amount of biofuel available would already be used in full to replace fossil fuels. Additional use of biofuels in, for example, a CHP plant would then have to be taken from another application. Subsequently, this other application has to resort to some other primary energy source, most probably a fossil fuel. A new biofuelled energy project causing additional biofuel consumption "on the margin" may thus lead to an increase in the use of fossil fuels somewhere else. To use the convention that biofuels emit no CO_2 could be regarded as misleading in such a situation, because it would imply that it does not matter how efficiently the biomass fuel is utilised. Grönkvist *et al.* [2003] defined "the Limited Biofuel Model" (LBM) where scarcity of biomass fuel is explicitly considered by debiting biomass fuel use with the same CO_2 emissions as coal. The direct comparison with coal comes naturally because coal can be replaced by biofuel in many applications without any extensive technical modifications. We applied the LBM, thus debiting additional biofuel consumed with 341 $kgCO_2$/MWh fuel. The effect of using this alternative approach is presented in the last columns of Tables 8 and 9. The results show that using the biomass in mills with CCS adds leverage to the CO_2 reduction potential of biomass, even when the limited biomass model is applied. Note that in this case emissions from the extraction of biomass were not accounted for because according to the LBM the biomass would be used for another application if it were not used for mill CHP requirements. This is an example of how scarcity of biomass could be included in an analysis based on the LBM. Region-specific conditions decide exactly at what rate biomass consumption should be debited with CO_2 emissions. Debiting could be based on another fossil fuel than coal, or a mixture of different fuels.

In conclusion, the results of Tables 6 - 9 show that substantial amounts of CO_2 could be captured in market pulp and integrated pulp and paper mills while biomass-based electricity could simultaneously be delivered to the grid on a net basis.[21]

---

[21] As a final note to the discussion concerning CO_2 balances we would like to add a comment on the post-combustion capture options. As previously mentioned an alternative option for CO_2 capture in pulp mills would be recovery boilers with post-combustion capture. For reference, applying the same methodology as for BLG-based systems, we made a simple estimation of CO_2 balances of CCS integrated with recovery boilers with back-pressure steam cycle. Steam data representing the most advanced recovery boilers in use today were assumed. CO_2 separation by chemical absorption using the chemical solvent Mono-Ethanolamine (MEA) was assumed. Energy required to regenerate the solvent and for stripping is typically 2.7-3.2 MJ/tCO_2 for state-of-the-art, high efficiency coal-fired power stations. As the CO_2 concentration of the flue gases from black-liquor and biomass-fired boilers are similar to that of a coal-fired boiler, steam consumption for regeneration was assumed to be 2,9 MJ/tCO_2 (MP steam). If additional biomass was regarded as CO_2-neutral, the results indicate that 10-40 % larger CO_2 reductions could be achieved with post-combustion option. This effect can be attributed to the inferior energy efficiency of systems based on recovery boilers in the following way; less efficient post-combustion technology leads to more additional biomass needed to satisfy the mill steam demand. As a consequence more CO_2 can be captured thus generating additional negative emissions. In this case it is illustrative to apply the limited biofuel model. If that is done, systems based on BLG are on par with, or slightly better than, the boiler based systems. This is explained by the superior energy efficiency of BECS based on BLG compared to BECS based on recovery boilers.

Cost Analysis

CHP System Capital Costs

Capital cost for the system components were first estimated by Möllersten et al. [2006]. The original cost data derives from several literature sources [Larson et al., 2000; STFI, 2000; Warnqvist, 2000; Brandberg et al., 2000; Williams, 2002; IEA, 2002; David and Herzog, 2000; Freund and Davison, 2002]. Tables 10 and 11 present the estimated capital costs. A scaling factor of 0.7 was used to adjust capital costs for size. An estimated initial accuracy of the source cost data is approximately 30%.

Table 10. Estimated capital costs of MPM/BLG CHP systems [MUSD]

Component	MPM/BLG$_1$	MPM/BLG$_2$	MPM/BLG$_3$
Black liquor gasification island	74	74	74
Biomass gasification island	-	-	-
Shift reactor	-	-	14
CO_2 absorbtion	-	7	14
Gas turbine	38	38	42[a]
HRSG	13	13	13
Steam turbine	7	6	4
CO_2 compressor	-	4	10
Total	**132**	**142**	**167**

[a]The turbine is fuelled with predominantly H_2. A 10% increase of the specific capital cost was assumed for the H_2-fuelled gas turbine.

Table 11. Estimated capital costs of IPPM/BLG CHP systems [MUSD]

Component	IPPM/BLG$_1$	IPPM/BLG$_2$	IPPM/BLG$_3$
Black liquor gasification island	74	74	74
Biomass gasification island	53	53	75
Shift reactor	-	-	20
CO_2 absorbtion	-	8	20
Gas turbine	47	47	57[a]
HRSG	16	16	18
Steam turbine	-	-	6
CO_2 compressor	-	5	13
Total	**190**	**203**	**283**

[a]The turbine is fuelled with predominantly H_2. A 10% increase of the specific capital cost was assumed for the H_2-fuelled gas turbine.

The Cost of CO_2 Transportation and Storage

The cost of CO_2 transportation and storage was determined using a model issued by the GHG Research and Development Programme of the International Energy Agency [IEA, 2002] (see model description earlier in this chapter). The assumed transportation cost as a function of transportation rate and distance is illustrated in Fig. 4. CO_2 injection was assumed to take place in CO_2-retaining deep saline formations with negligible seepage back to the atmosphere. [22] Pulp and paper mills are situated at sea or river ports enabling tanker transportation. Therefore, a cost ceiling of 20 USD/tCO_2 for transportation and storage was assumed representing the cost for long-distance tanker transportation (over 2000 km) [IPCC, 2005]. Longer transportation distances were regarded as unrealistic.

Modelling Framework

This section develops a model for evaluating the complex capital budgeting problem, which is analysed in this case study.[23] The model applied for optimising the pulp mill owner's decisions considers two real options. The first belongs to the category of capital options: an option to invest in building a new module (system). There are two possibilities concerning how the new module can be built. The owner can either build an entirely new module or, if at least one system has already been built, invest only in the components that are necessary to add to an already built module. The second option deals with the mill's operating strategy. Once a certain module has been built, we assume that the pulp mill owner has the flexibility to activate and deactivate it again if this behaviour is found to be profitable. We call this kind of option a "switch option". We will discuss these options in detail in the forthcoming paragraphs.

Actions

In the search for the optimal strategy of the pulp mill operation we consider a time horizon $T = 50$ years. In our model the pulp mill owner has an option to change the operation plan each year (that is 50 times altogether). The equipment used in the pulp or pulp and paper mill for the combined CHP/chemical recovery process (BLG island and supporting elements) reaches the end of its economic life time after $T_{retire} = 25$ years, that is there are at least two 25-year periods (called *CHP plant periods*) to be taken into account. However, if the pulp mill owner finds it optimal, the retirement of the equipment can be advanced, thus leading to a scenario with more than two CHP plant periods within the time horizon T.[24]

---

[22] Note that our cost assessment only considers dedicated single pipelines for each project. If a CO_2 grid with trunk pipelines would become a reality (similar to the case of natural gas) thus allowing numerous CO_2-emitting point sources to be connected to a CO_2 transport network, the average scale of the transportation system would increase thereby reducing the average cost per ton CO_2 transported. Under such circumstances, transportation by pipeline could be done over much longer distances before tanker transportation became the economically preferred option. It might be mentioned that the IEA (2002) estimated the cost of transporting CO_2 5000 km in large-diameter pipelines at 25 USD/tCO_2 (IEA, 2002), which gives some idea of the cost level that could be expected with a large-scale transportation system.

[23] For a comprehensive discussion on the model setup, please refer to Chladna et al. [2004].

[24] To abandon the current equipment before the actual retirement time elapses could be a reasonable decision if, e.g., the prices that the pulp mill owner takes into consideration change suddenly and the equipment is near the end of its economic life time. Then the owner faces the following three options in our model: 1. To ignore the sudden change in prices, wait until the equipment reaches the end of its economic life time and then invest in new equipment; 2. To add to the existing equipment immediately; such investments must, however, be fully renewed in a few years when the original equipment reaches the end of its economic life time; 3. The existing

There are three types of BLG-based modules (denoted by BLG_1, BLG_2, and BLG_3), which we are interested in. Since exactly one module must be running in each year, the pulp mill owner deals with the following options every year:

- To invest in and build a new module, which has not yet been built in the current CHP plant period. This option will be called a *capital investment action*. Note that this action is a necessary choice in the first year of the respective period;
- To switch to using a module that has been built earlier in the current CHP plant period, but then (as perhaps temporarily economically inefficient) deactivated. This, in fact, means that an existing module will be re-activated. This option will be called a *switch action*; and
- To stay with the module that has been active so far. This option will be called a *stay action*.[25]

We assume that each BLG module is actually organized into components, which can be reused. For example, if the pulp mill exploited the module BLG_1 so far and the owner now decides to upgrade to the BLG_3 module, the components that the modules BLG_1 and BLG_3 share in common have not to be built: the respective components from BLG_1 can be reused within BLG_3. However, if the owner terminates the current period (either because the retirement time T_{retire} elapses or because she just finds it optimal), no reuse is possible anymore; all components must be built again from the beginning (and can be used or reused for the next T_{retire} years).

Costs

Naturally, taking each of the above actions induces costs that we must consider. Therefore, we assume that there is a cost function $c(.)$, set up as an input parameter of the model that prescribes the necessary investment for the respective actions:

- $c(m_1 | M)$: the cost of a capital investment action. With this action we build a new module m_1 assuming that each module in the set M has already been built earlier in the current CHP plant period. Therefore, the components of modules in M can now be reused. The module m_1 requires components of two kinds: the units that are not present in any module in M (i.e., the components that must be constructed for the first time in the current CHP plant period) and the components that are contained in some module of M (i.e., the units that will be reused). The cost $c(m_1 | M)$ (also known as *capital investment cost*) comprises not only the investment in the former units but also the costs associated with repetitive implementation of latter ones;
- $c(m_1 \rightarrow m_2)$: the cost of a switch action, when switching from using module m_1 to using the existing module m_2;[26]

---

equipment can be retired prematurely and investment made immediately in equipment upgrade (that is, the owner has an option to enforce the start of a new CHP plant period). When compared to option (2), such investment need not be renewed in a few years (which is an advantage), albeit there are losses due to premature retirement of the old equipment (which is a disadvantage). It is reasonable that, in certain cases, the third option becomes superior which justifies our decision to consider it.

[25] Performing two actions in the same year is not allowed however.

- $c(stay) = 0$: the cost of the stay action, as it requires no investments.

As there are three BLG-based types of modules, only 6 values must be specified to cover all possible switch actions. Similarly, only 12 values must be specified to cover all possible capital investment actions. Note that these costs are not the only ones associated with the module operation (see, for example, operational costs, transportation costs, etc., which we introduce later).

We assume that the structure of function c is "reasonable". In particular, we assume that:

- Switching to the usage of an existing module cannot be more expensive than building that module for the first time, under otherwise identical circumstances, in the same year (that is, re-activation of a module is a reasonable choice);
- Capital investment is not made into a module if the same module was built earlier in the same CHP plant period (that is, the reactivation of a module component is always cheaper than its construction "from scratch").

Learning

In our model we consider technological learning by introducing the learning rate R. This means that any capital investment cost or switch cost, or any other technology-related cost has to be decreased by the factor $1/(1 + R)^t$, if associated with year t.

Note, that there is a significant difference between the effect of learning and the effect of discounting: Learning means real reduction in costs, while discounting is just projection of the same amount of money in time. The time 0 cost of investment c performed in the year t is therefore:

$$\frac{c}{(1+r)^t (1+R)^t}$$

that is, both effects apply.

Price Processes

So far we have considered the module set up. However, once the module is active it operates and perhaps produces profit (i.e., operational profit). In our model, the operational profit is governed by price processes varying in time: the electricity price p_t^e, the biomass price p_t^b, and the CO_2 price p_t^c. This introduces uncertainty into the model as the above prices are generated/simulated (see the subsequent section on generating price processes for details of the price processes simulation).

Operational Profit

When the module m operates, it produces a fixed amount $q_c(m)$ of captured CO_2 and surplus electric power, $q_e(m)$, on an annual basis. It is assumed that the captured CO_2 is sold at the mill fence to an external undertaker that carries out the transporting and storage activities

---

[26] The switching costs were defined to be 15% of the capital cost for components that are switched on from an off-state and 10% of the capital costs for components that are switched off into off-state.

and, furthermore, that the mill owner and the CO_2 undertaker share equally the revenue which is the income from selling CO_2 emission permits reduced by the cost for transportation and storage. Depending on prices, the income for the pulp mill owner in the year t is calculated as follows:

$$0.5 \times p_t^c \times q_c(m) \; + \; p_t^e \times q_e(m).$$ (1)

The associated production costs in the year t are:

$$p_t^b \times q_b(m) \; + \; \frac{c^{oper}(m)}{(1+R)^t} \; + \; \frac{0.5 \times c^{trans}(m,d) \times q_c(m)}{(1+R)^t},$$ (2)

where q_b denotes the biomass requirement (additional to black liquor) for the CHP process and c^{oper} is the annual operational cost related to production. Furthermore, we assume the fixed annual transportation and storage costs (c^{trans}) for each unit of CO_2 captured given the transportation distance d. The difference of equations (1) and (2) will be called operational profit $p_t^{oper}(m)$.

The values q_c, q_e, q_b, c^{oper}, c^{trans}, and d are input constants of the model. We will assume that the annual costs for operation of module m amounts to 4% of its full initial capital investment costs. Note that the last period may be subject to different treatment as is outlined below.

Special Case: The Last CHP Plant Period

The last CHP plant period requires special treatment in computations, as it may be artificially and thus prematurely terminated when the simulation reaches the time horizon T. In such a case the capital investment originally meant for period of T_{retire} years is applied in an artificially shortened period, thus the corresponding action could be found to be suboptimal. Instead of approximating the future profit that one would obtain if operating the module behind the time horizon T (which is hard to estimate), we reduce the capital investment costs in the last CHP plant period by a factor relative to the length of the last period with respect to the full T_{retire}-long period. (For example, if the last CHP plant period is artificially terminated by a time horizon T after five years, all capital investment costs realized within this period are reduced by factor 5/25 = 20%.)

Solution Techniques

The presented study is an application of the real option theory,[27] which is nowadays a widely accepted technique in the analysis of investment problems under uncertainty.

In our analysis we assume that the uncertainty arises only from the fluctuation of the prices in time. Therefore we assume that the stochastic part of the operational profit corresponds to the underlying source of randomness (S_t). Further, we assume that it can be

---

[27] See, e.g., Dixit and Pindyck, (1994) - an excellent overview of the real options techniques

modeled as the geometric Brownian motion process[28], which is a classical assumption in the real option analysis. Actually, we adopt the real option framework in discrete time and determine the optimal actions by building a decision tree.

A first step in the real options analysis requires us to construct a binomial lattice, which represents the time development of the underlying process. A basic idea of the binomial lattice construction is to structure the stochastic process in the simple manner: it is assumed that the underlying process can either tick up or tick down at each time point. Further, the discretization is proposed in such a way that the number of time steps is sufficiently large in order to achieve the same results as in the continuous case.

Once the binomial lattice has been constructed the optimal decisions can be determined for each decision node. We use the standard backward moving algorithm to solve the problem.

At each node for which a feasible decision can be adopted, the optimal decision is given as a solution of the following optimization problem:

$$V(\omega,t) = \max_{a_t} \pi(S_t,a_t) + e^{-r\Delta t} E(V(f(\omega,a_t),t+\Delta t)), \qquad (3)$$

i.e., the maximization procedure selects that feasible action, which maximizes the sum of the current profit flow π plus the expected discounted value of the continuation profit. In Equation (3) S_t represents the underlying process, a_t the feasible action, and the symbol E stays for the expected value operator. For the real option analysis we assume that a sufficiently wide set of financial instrument exists, such that the standard replication argument known from the option theory can be applied. Under such a consideration expected value in the formula (3) is discounted using the risk free rate r. This also means, that the expected value is determined using the risk neutral probabilities with respect to the underlying binomial lattice.

Finally, note that due to the structure of the problem space, the states ω, that is the nodes of our binomial lattice, are actually multidimensional, therefore optimization described by (3) is multidimensional as well: taking the action a_t corresponds hence to the change of state described by the transition function f.

Generating the Price Processes

Price information has been taken from Riahi et al. [2003] and Nakicenovic and Riahi [2002]. Riahi et al. [2003] computed electricity and carbon prices as shadow prices in GHG stabilisation runs including CO_2 capture technologies aiming at an atmospheric CO_2 concentration of about 550 ppmv. Two stabilisation scenarios for each baseline were developed — one assuming constant costs for capture technologies (A2-550s, B2-550s), and one including learning for capture technologies (A2-550t, B2-550t). All four stabilisation scenarios are based on iterated runs of the global optimisation framework MESSAGE-MACRO [Messner and Schrattenholzer, 2000]. Global emissions peak at about 9 to 12 GtC around 2050 and then proceed to decline to slightly less than the 1990 emissions level (6 GtC) by 2100. These emissions profiles are similar to other emissions trajectories for 550

---

[28] e.g., we assume that the underlying process can be modelled as: $dS_t = \mu_t dS_t + \sigma_t dW_t$, where dW is an increment of the Wiener process. For more details see Dixit and Pindyck, (1994), page 68.

ppmv stabilisation cases found in the literature (for example, Wigley et al., 1996; Riahi and Roehrl, 2000).

CO$_2$ Price

The carbon value is an endogenous output calculated by the MESSAGE model. It can be interpreted either as a carbon tax or value of an emission permit that has to be introduced in a GHG-constrained world in order to meet the stabilisation target. In the stabilisation scenarios, CO$_2$ prices grow steadily from about 5 US$/tCO$_2$ in 2020, to about 7–17 US$/tCO$_2$ in 2050, to about 33–70 US$/tCO$_2$ in 2070, and to about 110–136 US$/tCO$_2$ in 2100. The sharp increase at the end of the century is partly due to discounting with a 5% annual rate. In order to constrain the range within which the prices are allowed to fluctuate we selected the A2-550t price trajectory as an approximation of the upper boundary and the B2-550t price trajectory as an approximation of the lower boundary. Mainly due to discounting, but also due to population and GDP growth the CO$_2$ price process is modelled to be time dependent. More precisely, the CO$_2$ price trajectories are generated as follows:

$$p_t^c = e^{c_1 + c_2 t} + \varepsilon_t^c, \qquad (4)$$

where $\varepsilon_t^c \sim N(0, (\sigma_t^c)^2)$. The parameters c_i, $i = 1, 2$ were estimated to fit the carbon shadow prices by Riahi et al. [2003] as described above. σ_t^c were conjectured based on subjective judgment.

Figure 6 shows the CO$_2$ price trajectories produced by Equation (4) for 10 simulations.

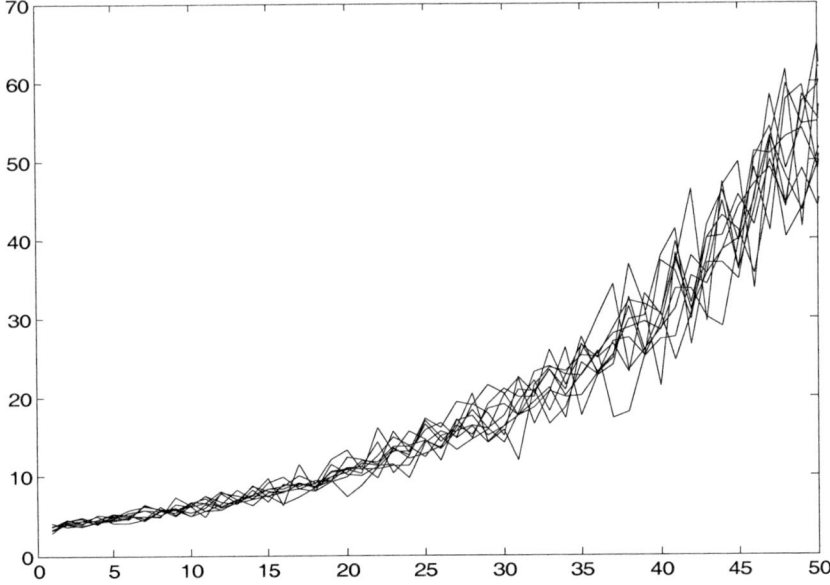

Figure 6. CO$_2$ price trajectories produced by 10 simulations.

Electricity Price

Riahi et al. (2003) compute an electricity price increase of about 100% in the coming century due to the stabilisation constraint. Based on the expectation of the price increase we fitted the electricity price trajectory to the following equation:

$$p_t^e = e^{e_1 + e_2 t} + \varepsilon_t^e \tag{5}$$

where $\varepsilon_t^e \sim N(0, \sigma_e^2)$. The electricity price is modelled as a time dependent process. In fact, we use the same structure as in case of the CO_2 price.

In order to obtain the underlying data for the parameter estimation of the process (5) we altered the electricity shadow prices in OECD countries calculated by Riahi et al. (2004) such that the particular information from the Nordpool energy market is taken into the account. We see such an adjustment of the global OECD scenario as a necessary modification in order to predict the prices in particular countries. This is due to the fact that electricity cannot be freely traded among the countries (for example, because of limitations in transmission capacities.) More precisely, as the main underlying information for the parameters estimation we consider the historical evolution of the yearly electricity spot price averages in Sweden modified by trends of the electricity price in the OECD countries predicted by the MESSAGE model. σ_t^e and the correlation between the CO_2 and the electricity price, $\rho(p^c, p^e)$, were conjectured based on subjective judgement. Figure 7 shows the electricity price trajectories produced by Equation (5) for 10 simulations.

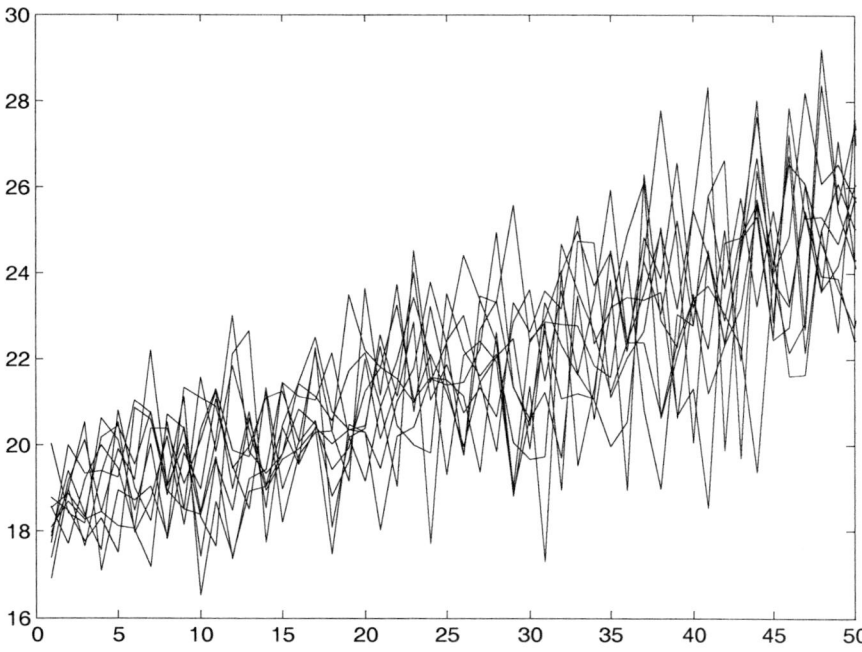

Figure 7. Electricity price trajectories produced by 10 simulations.

Biomass Price

The modelling of the biomass price follows on the results presented by Ådahl and Harvey [2004]. In this study the authors propose a price-setting model for biofuel that assumes a constant price ratio for biomass and electricity in the Nordic countries. Empirical evidence shows that the value of the ratio is close to 2.9. However, so far, biomass fuel markets have not faced the impact of regional and global biomass fuel trading. Biomass markets might become increasingly international with global trade, main importers of biomass being the OECD countries.

Therefore we suggest that the development of the biomass price in Sweden will be correlated to the average of Sweden *and* the OECD marginal electricity prices. The biomass prices are generated as:

$$p_t^b = e^{b_1 + b_2 t} + \varepsilon_t^b, \tag{6}$$

where $\varepsilon_t^b \sim N(0, (\sigma_t^b)^2)$. Similarly to the electricity price and the CO_2 price, the biomass price is hence modelled as a time dependent process. The parameters b_i, $i = 1, 2$ were estimated to fit the biomass prices generated by the procedure proposed above. σ_t^b and the correlation between the CO_2 price and the biomass price, $\rho(p^c, p^b)$, were conjectured based on subjective judgment.

Figure 8 shows the biomass price trajectories produced by Equation (6) for 10 simulations.

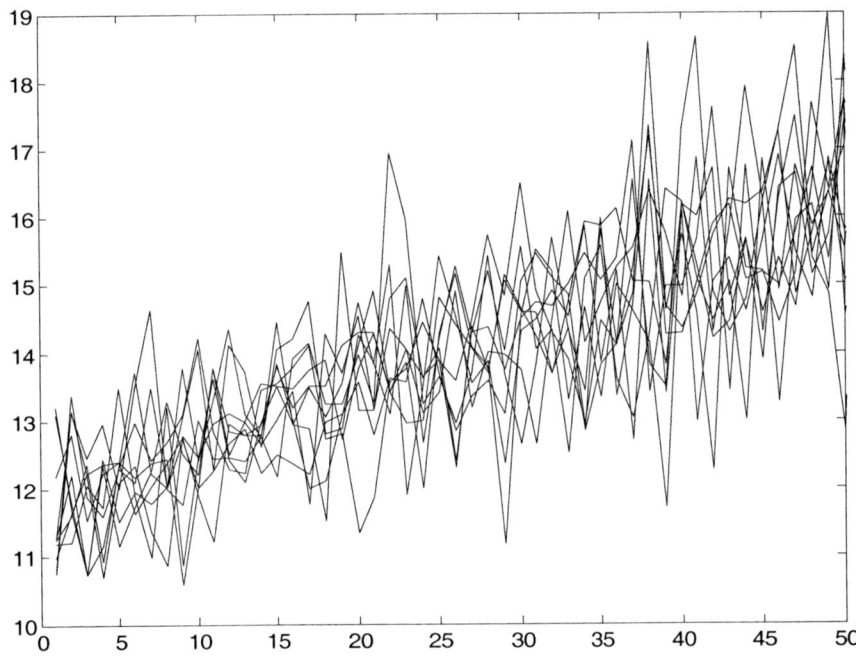

Figure 8. Biomass price trajectories produced by 10 simulations.

Results of the Economic Evaluation

In order to analyze the potential of commercial use of the described MPM/BLG and IPPM/BLG modules, respectively, we have performed several numerical experiments. We have simulated the price trajectories and used the decision tree (see Section "Solution Techniques") to find the optimal sequence of a pulp mill owner's actions. We have run 1000 simulations and considered the total time period to be 50 years ($T = 50$) starting in 2020^{29}, the retirement time of the equipment to be 25 years ($T_{retire} = 25$) and the risk free rate (r) to be 5%. Since our main interest is to find the expected optimal time to enter the carbon market, i.e., to start capturing CO_2, the presentation of the results will be biased towards CO_2 capture technology. The results will be presented so that the sensitivity of optimal commitment to technological learning assumptions and various transportation distances for CO_2 can be assessed. We consider three different learning rates: 5%, 10% and 15%, respectively and three different CO_2 transportation distances: 100 km, 400 km and 1000 km, respectively. It means that altogether we present the results of nine numerical experiments in each MPM/BLG and IPPM/BLG case.

Market Pulp Mill

When dealing with MPM/BLG modules, the numerical experiments show that BLG_1 is consistently chosen as the optimal module to be built in the first year. Module BLG_3 is built in all nine cases, too. The building time of BLG_3 is concentrated within the first eleven years of the second simulation period (i.e., within years 26-36). However, it is observed that BLG_2 is not a competitive technology: in most cases we observe a direct switch from BLG_1 to BLG_3. Only in some cases BLG_2 acts as a transitory technology. This phenomenon is significant only under the high learning rate. Simulations show that under the high learning rate in approximately 53-71% of simulations BLG_2 is built in year 26 that is in the first year of the second simulation period.

Another interesting aspect is the sensitivity analysis of the expected commitment time for BLG_3 technology. Figure 9 presents the resulting frequency distributions with the emphasis on the effect of the assumed learning rate. Not surprisingly, irrespective of the length of the CO_2 transportation distance, the higher the learning rate, the sooner the first introduction of BLG_3 technology.

---

[29] The starting year 2020 has been chosen because the IPCC [2001b] predicts that CCS could give major contributions to CO_2 abatement by 2020. It may, therefore, be reasonable to assume the existence of a readiness (technical and institutional) to implement CCS by that year.

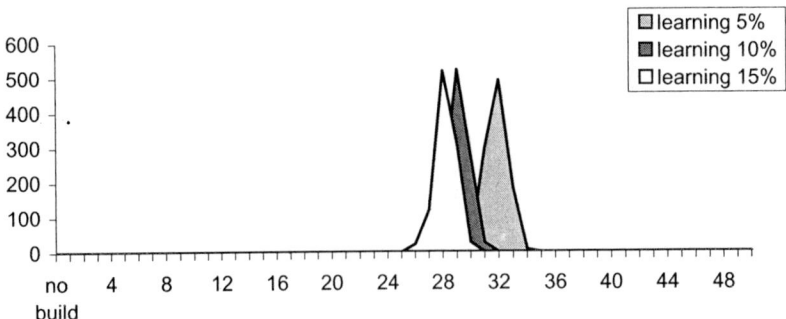

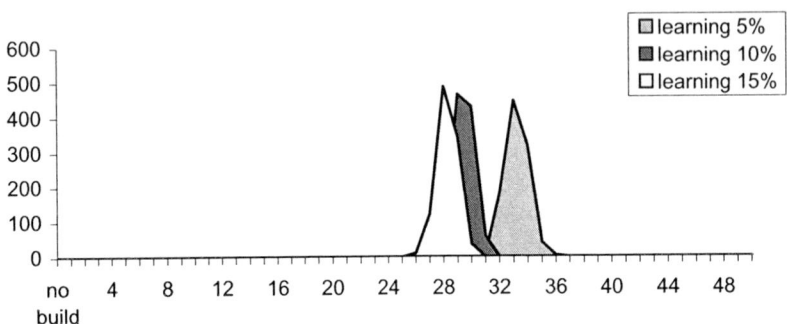

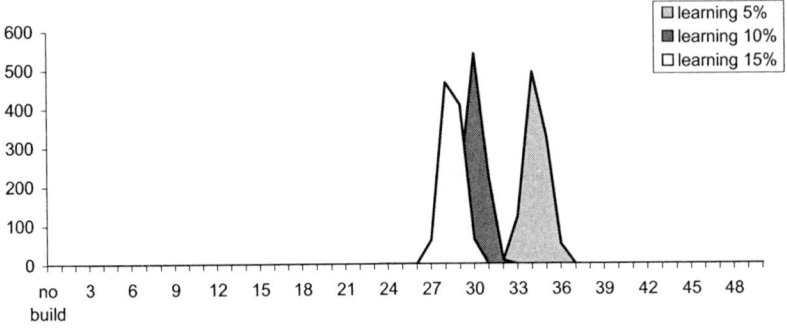

Figure 9. Frequency distribution of commitment time for MPM/BLG$_3$ technology.

The spread of BLG$_3$ commitment times caused by change in the learning rate varies from about 3.6 years for the case of short transportation distances (100 km) up to 5.9 years in the case of long transportation distance (1000 km). Sensitivity analysis with respect to CO$_2$ transportation distance shows overall less variability. The results in form of expected commitment time are depicted in Figure 10. We observe that the transportation distance does not influence the expected commitment time significantly: the maximum[30] spread induced by the increase of the CO_2 transportation distance (from 100km to 1000km) is 2.4 years.

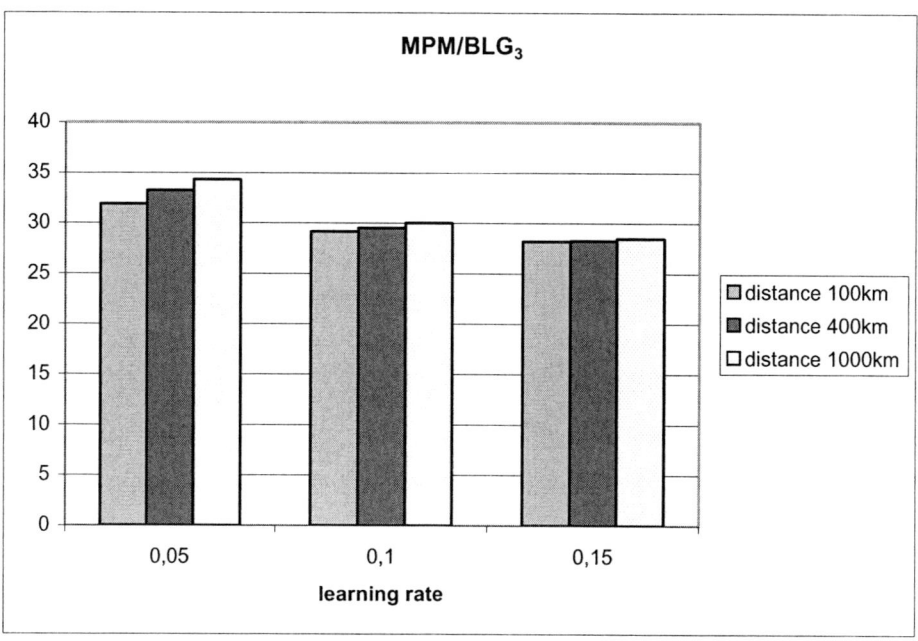

Figure 10. The expected commitment time for MPM/BLG_3 technologies under different transportation distances subject to learning rate.

Integrated Pulp and Paper Mill

Similarly to the MPM/BLG case, the module BLG_1 is always built in the first year of the first investment period. However, in the IPPM case, BLG_2 acts as an active transitory technology: no direct transition from BLG_1 to BLG_3 has been observed. Actually, in all simulations the modules have been built sequentially in the following order: first BLG_1, then BLG_2 and only afterwards BLG_3. The exact timing of the occasion that BLG_2 and BLG_3 are built, respectively, depends on the learning rate and transportation distance setup.

Figure 11 shows that the expected time of the first commitment of BLG_2 varies between years 18 and 28. Not surprisingly, the earliest commitment is observed under the fastest learning (15%) and the shortest CO_2 transportation distance (100 km). For higher learning rates (10% and 15%) the technology switch occurs always prior to first technical retirement and, most interestingly, earlier than in the MPM case.

---

[30] Reached under the learning rate (5%).

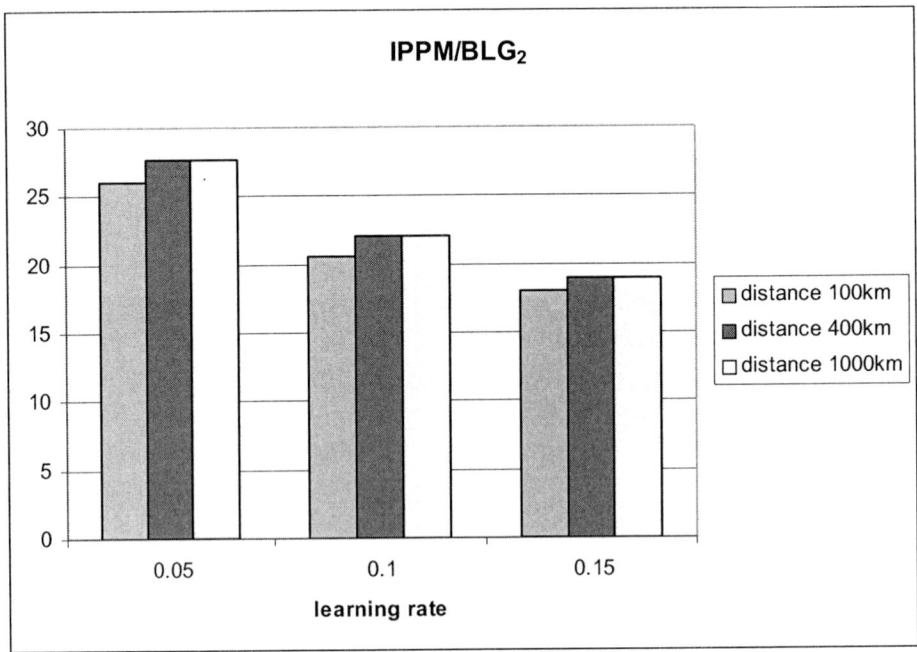

Figure 11. The expected commitment time for IPPM/BLG$_2$ technologies under different transportation distances subject to learning rate.

In contrast to this, for the slow learning assumption (5%), irrespective of the length of the transportation distance, the expected building time of BLG$_2$ is concentrated within the first three years of the second investment period (26-28). Furthermore, we observe that for the slow learning rate (5%) and short transportation distance (100 km), the timing to build BLG$_2$ is mostly dominated by the technical retirement restriction (i.e. lifetime of 25 years): the first build of BLG$_2$ is postponed to the second investment period rather than being switched to just before the end of the first period.

Figure 12 presents the frequency distribution of the first commitment time of BLG$_3$ module.

The numerical results show that BLG$_3$ is always committed after the first technical retirement: for all setup combinations of learning rate and transportation distance the optimal building time falls within the range of years 27-35.

As we already mentioned, under all nine studied setups BLG$_3$ is built subsequent to the BLG$_2$ module. We calculated the transit period to be approximately 6-9 years. A variance of the first commitment time of BLG$_3$ is sensitive neither to the transportation distance nor to the learning rate. However, the variance is greater than in the MPM/BLG$_3$ case.

Moreover, Figure 11 and Figure 13 group the expected time of the first commitment of the BLG$_2$ and BLG$_3$ so that the sensitivity with respect to the CO$_2$ transportation distance can be easily derived. The results show that in both cases commitment times are rather insensitive to the transportation distance of CO$_2$ — despite the rather high assumption on the transportation price of about 20 €/tCO$_2$ for longer distances. The maximum spread of the expected commitment times caused by the change in the transportation distance is only 1.65 years.[31]

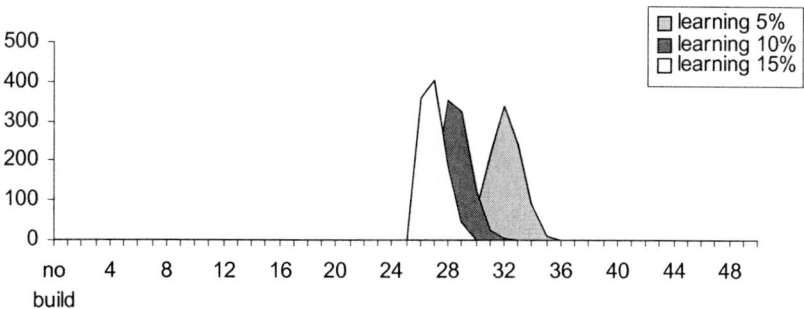

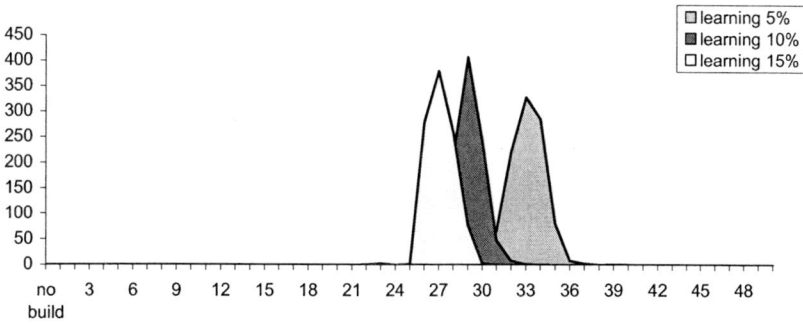

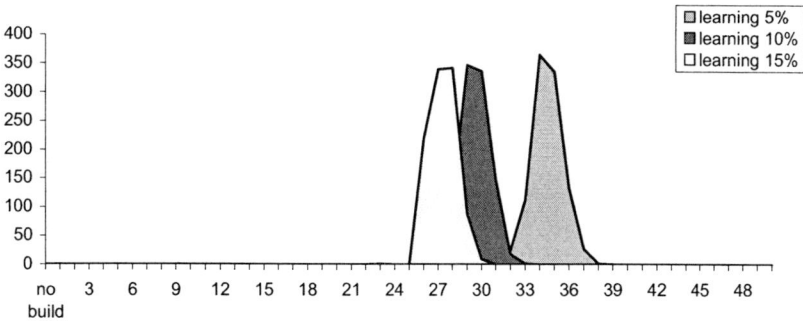

Figure 12. Frequency distribution of commitment time for IPPM/BLG$_3$ technology

---

[31] Considering the positive economies-of-scale in CO_2 transportation, the results' insensitive character (for the MPM and IPPM cases) to the distance from the point of capture to the point of storage comes with some surprise since the analysed biomass-based energy conversion systems are of modest scale with associated modest rates of CO_2 generated.

Moreover, Figure 11 and Figure 13 group the expected time of the first commitment of the BLG_2 and BLG_3 so that the sensitivity with respect to the CO_2 transportation distance can be easily derived. The results show that in both cases commitment times are rather insensitive to the transportation distance of CO_2 — despite the rather high assumption on the transportation price of about 20 €/tCO_2 for longer distances. The maximum spread of the expected commitment times caused by the change in the transportation distance is only 1.65 years.[32]

On the other hand, the learning assumptions make a notable difference: in general, the higher the learning rate, the sooner the introduction of CCS. The spread caused by the increase of the learning rate varies from 8 up to 8.6 in case of BLG_2 and from 5.1 up to 7.2 years in BLG_3 case, respectively.

Overall, the sensitivity analysis of the MPM/BLG and IPPM/BLG results feature similar characteristics.

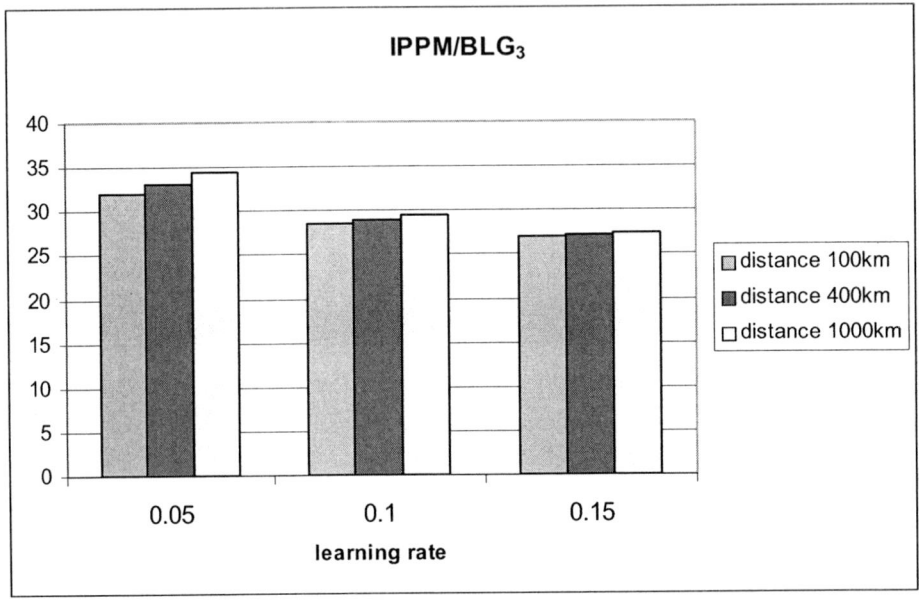

Figure 13. The expected commitment time for IPPM/BLG_3 technologies under different transportation distances subject to learning rate.

---

[32] Considering the positive economies-of-scale in CO_2 transportation, the results' insensitive character (for the MPM and IPPM cases) to the distance from the point of capture to the point of storage comes with some surprise since the analysed biomass-based energy conversion systems are of modest scale with associated modest rates of CO_2 generated.

DISCUSSION AND OUTLOOK ON NEGATIVE EMISSION BIOMASS TECHNOLOGIES

In this chapter we have evaluated investments in BECS in a pulp and paper mill context. The analysis considered biomass-based systems with a fuel input between approximately 300 to 500 MW. Our analysis shows that in the range of scales considered, BECS can be economically feasible within approximately 40 years given carbon, electricity and biomass prices predicted by one of the leading IPCC global energy scenario models.[33] However, it has to be noted that market imperfections may yield much higher prices than those predicted in an idealized model world, which would suggest that price signals triggering BECS could emerge earlier.

As mentioned in the introduction to this chapter, an overwhelming majority of all work done related to CCS has focused on applications to emissions from fossil fuels. One likely explanation is the issue of scale. Sizes of coal-fired power plants in the range from 500 to over 1000 MWe are currently common in the world and natural gas-fired power plants with unit sizes of several hundred MWe are abundant. Pulp and paper mills, along with sugar cane-based ethanol mills, are among the few industries where large-scale bioenergy conversion takes place in the world today, thus providing potential niche markets for BECS implementation [Möllersten et al., 2003a].[34] The fact remains, however, that most of the biomass in the world is currently used in traditional ways at small scales not compatible with CCS for economic reasons. Nevertheless, the IPCC [2005] states that "it is perfectly conceivable that these technologies [BECS, authors' remark] might play a significant role by 2050 and produce negative emissions across the full technology chain". This would require a significant increase in large-scale biomass energy conversion on a global scale compared to the present situation. In the future, one might expect small-scale uses of biomass to continue to remain significant, but larger-sized operations compatible with CCS is likely to become increasingly important with increasing interest in bioenergy implementation. In order to justify this statement we refer to Table 12 that summarises the results from several studies concerning economically optimal scales of bioenergy conversion. Given the rather large optimal scales, a growth in commercial bioenergy implementation could very well bring about an increase in large-scale bioenergy conversion, which would imply improved opportunities for BECS.

---

[33] In the pulp and paper mills investigated in this chapter approximately half of the biomass delivered to the mill ends up as fuel and the other half as fibre. Moreover, large chemical pulp mills of today produce in the range 1500–3000 tons of pulp per day, i.e. up to twice as much as the mills studied here. Thus, it has already been proven that it is logistically feasible to operate facilities that receive biomass at very high rates - in the order of 1000 MW.

[34] In addition, in some countries (e.g. Sweden and Finland) large heating and CHP plants connected to district heating networks are fired with biomass.

Table 12. Estimates of optimum scales of bioenergy conversion[a]

Type of plant/mode(s) of operation	Fuel supply	Optimum scale [MW]	Author(s)
BIG/CC/Power production	Dedicated energy crop (switchgrass) or plantation wood	230-320 MWe	Larson and Marrison [1997]
BIG/CC/Power production	Dedicated energy crop (switchgrass) or plantation wood	110-142 MWe	Larson and Marrison [1997]
BIG/CC/Power production	Dedicated energy crops (williow)/Municipal RDF	COE declined with increasing scale within the studied capacity range from 51-215 MWe	Faij et al. [1998]
Fluidised bed combustion and grate firing/ Power production and CHP	Forestry residues/industrial waste wood	Up-scaling decreases costs per unit primary energy saved within the studied range 10-200 MW thermal input	Dornburg and Faaij [2001]
Fluidised bed gasification/ Power production and CHP	Forestry residues/industrial waste wood	Up-scaling decreases costs per unit primary energy saved within the studied range 10-300 MW thermal input	Dornburg and Faaij [2001]
Steam turbine power plants/ Power production	Forestry residues/ agricultural residues/whole forest	137/450/900 MWe	Kumar et al. [2002]
Bioethanol plants	Cane sugar/Cane sugar plus sorghum	140-200 MW thermal output	Nguyen and Prince [1996]
Electricity production/ Synthetic fuel production	Dedicated energy crops	Approximately 1000 Mw thermal input	Greene [2004]
Methanol, ethanol, hydrogen and Fischer-Tropsch diesel plants	Woody biomass	Strong economies-of-scale effects between 80 and 400 MW thermal input. Production costs continue to decrease between 400 and 2000 MW thermal input albeit at a slower rate.	Hamelinck [2004]

[a]All studies presented in the table consider scale effects in the cost of biomass feedstock supply. All authors except Hamenlinck [2004] assumed the biomass to be produced near the conversion facility. Hamenlinck [2004] considered long-distance international biomass transportation from biomass production areas to energy import regions. The analysis, therefore, assumes a flat biomass feedstock cost. The investigation shows that transcontinental bioenergy trade is possible while maintaining competitive biomass prices and modest energy losses. Hamenlinck [2004] concludes that the energy requirement to deliver solid biomass from South America to Europe is 1.2 – 1.3 MJprimary/MJdelivered. As a comparison, the corresponding energy requirement for coal is approximately 1.1 MJ/MJ.

The potential for converting biomass into long-term carbon-sequestering charcoal (see the introduction to this chapter) adds to the total theoretical potential of negative emission biomass technologies. Whereas BECS is a centralised technology, charcoal carbon sequestration is feasible for small-scale distributed biomass based energy generation and carbon sequestration. In this sense the two technology groups are complementary. It is also worth noting that while BECS requires advanced technologies and technological know-how, charcoal carbon sequestration can be based on relatively simple and well-known technologies and practices.

Taking these considerations into account, it is perceivable that negative emission biomass technologies could turn out to be an important option in a wider portfolio of GHG abatement technologies allowing for long-term energy security and sustainable climate management. The uncertainties surrounding the competitiveness of negative emission biomass technologies is however larger than that of conventional mitigation technologies Therefore, we will conclude with a discussion concerning areas where Research and Development could fill important knowledge gaps. Note that the discussion that follows focuses on issues specific to negative emission biomass technologies. Research and Development in the general area of CCS and energy technology, such as methods for biomass production, technologies for transportation and storage of CO_2, issues surrounding the integrity of CO_2 storage options, CO_2 separation technologies, thermochemical conversion of solid fuels and advanced bioenergy conversion will be beneficial also for the development of negative emission biomass technologies.

Important Areas of Research

Technologies and Systems

BECS

Although a broad variety BECS "technology chains" can be envisaged, linking different biomass feedstock sources, biomass pre-treatment and transportation systems, bioenergy conversion and CCS options, only few BECS "technology chains" have been studied in any detail.

Regional solutions that encompass biomass production, bioenergy conversion and CCS in one region are one possibility, but another option would be to integrate long-distance transportation of solid biomass or refined fuels at various stages along the technology chains. For example, BECS could involve large-scale production of transport fuel with CCS in biomass-producing regions and subsequent export of the fuels produced to energy-importing regions. BECS could also comprise biofuel export from biomass-producing regions to large-scale energy conversion facilities in energy-importing regions where CCS would then take place. One further option would be to combine the latter two examples.

A number of competing bioenergy technologies will likely be available. Today's biomass to electricity or CHP capacity is based on mature, direct-combustion boiler/steam turbine technology. Direct-fired combustion technologies for electricity or CHP production can be integrated with post-combustion or oxy-fuel CO_2 capture. An important near-term low-cost option for the extended use of biomass is co-firing with coal in existing boilers, i.e. the practice of introducing biomass as a supplementary fuel in high efficiency boilers (see, for example, Veijonen et al, 2003). Due to scale advantages, co-firing is likely to be a cost-effective BECS option in the more near-term [IPCC, 2005]. In addition, the flexibility to choose between fossil and a carbon neutral feedstock allows to hedge against uncertainties of future carbon market uncertainties. Another potentially attractive biomass to power or CHP option is based on gasification. Biomass integrated gasification/combined cycle (BIG/CC) systems would be expected to have thermal efficiencies nearly double those of direct-combustion systems. Biomass gasification is, as mentioned earlier in this chapter, compatible with pre-combustion CO_2 capture technology. Advanced biomass power systems based on gasification benefit from the substantial investments made in coal-based gasification combined cycle systems. Biomass gasification systems with CO_2 capture will also be appropriate to provide fuel to fuel cell and hybrid fuel-cell/gas-turbine systems. A near-term opportunity for BIG/CC technology is in the forest products industry (more details are presented in the main section of this chapter).

The few published studies concerning BECS in the power sector include Audus and Freund [2004], Hochenauer et al. [2004], Larson et al. [2005] and Rhodes and Keith [2005]. Audus and Freund [2004] identified high costs for CCS with a 30 MWe BIG/CC power plant. CCS doubled the COE compared to a case without CO_2 capture. The study clearly illustrates the high cost of CCS in small-scale applications. With respect to CHP several studies have been carried out analysing the energetic performance, carbon balances and costs of BECS in pulp and paper mills, considering several system configurations including post-combustion capture, pre-combustion capture and oxygen combustion. The main section of this chapter summarises studies in this area.

Biomass-based liquid and gaseous fuels ("biofuels") can be produced either via biochemical processes or gasification. Gas emissions from fermentation, the biochemical process that converts sugars into ethanol, are concentrated CO_2 and capture of CO_2 is essentially an activity consisting of condensing-out water and compression. Methanol, hydrogen and so-called Fischer-Tropsch diesel are examples of biofuels that can be produced from biomass via gasification. Several production routes are possible. CO_2 separation is part of all processes, which reduces the additional cost for employment of CCS. With respect to liquid fuels and hydrogen, Walsh [1993], Williams [1998] and Möllersten and Yan [2001], Larson et al. [2005] presented studies investigating BECS-based production of methanol, hydrogen, and Fischer-Tropsch fuels.

In addition to the publications mentioned above, Azar et al. [2006] estimate key performance figures of different BECS alternatives calculated on a consistent basis. Capital costs and energetic performance of biomass-based power, CHP, liquid fuel and hydrogen production with CCS are estimated, albeit on a highly aggregated level.

It is evident that these BECS studies provide very limited information regarding the potential of BECS in the power sector, for CHP outside the pulp and paper sector as well as in relation to the full range of biofuel production options. The efficiency and economic performance of BECS needs to be studied over the full range of fuel supply options, bioenergy conversion technologies and CO_2 capture methods. The potential improvement through integrated process configurations and the development of new technologies needs to be studied further. Furthermore, a geographically explicit analysis of the relationship between possible sources and storage locations would add valuable information. Optimal scales for BECS have not been analysed. Note that the estimated optimum facility sizes for bioenergy conversion quoted in Table 12 are for applications without CCS. Because of positive economies-of-scale in CCS, combining bioenergy conversion with CO_2 capture could be expected to lead to elevated optimum sizes. This issue would need to be further analysed.

Charcoal Carbon Sequestration

Charcoal carbon sequestration technologies are based on the carbonisation of biomass materials to make charcoal which is subsequently put into repository or utilised for carbon-sequestering purposes such as soil amendment in agriculture and forestry or even advanced building material. In its least advanced form, the simple transformation of organic carbon in biomass into inorganic carbon in the form of wood charcoal in some common type of furnace, the carbonisation process involves technologies that are readily available. With modern technologies hydrogen can be co-produced with charcoal using reforming technologies. The hydrogen produced can be converted to synthetic liquid fuels or used in, for example, fuel cells..The charcoal can be further processed for its use as a soil amendment.

Okimori et al. [2003] presented a case study based on carbonization of biomass residue and waste from tree plantations and pulp mills. Ogawa et al. [2005] analysed three cases to evaluate the potential of the charcoal carbon sequestration finding fixed carbon recovery ratios[36] in the range 21 to 32 %. In a study by Day et al. [2005] a process has been proposed which co-produces a nitrogen-rich, slow-release charcoal fertiliser and hydrogen. Furthermore, the combination of the nitrogen-enriched charcoal production process with a chemical process that can directly convert CO_2, SO_x and NO_x from fossil fuel plants to valuable fertilizers is proposed.

The number of studies analysing process solutions is quite limited, which calls for futher Research and Development. Further work is also needed to reduce the uncertainty surrounding the stability of charcoal in soil. Ogawa et al. [2005] point out that there are only a few studies that clarify oxidation and degradation processes of carbon in charcoal.

Monitoring, Inventories and Accounting

The requirement for explicit treatment of BECS with respect to accounting issues has been noted in some publications [IEA, 2004; IPCC, 2005; Grönkvist et al., 2006]. Grönkvist et al. [2006] propose desirable characteristics for an accounting approach for BECS and suggest that the desirable criteria can be achieved with an accounting approach that would allow for the assignment of "removal units" in a biotic CCS carbon pool for biomass CO_2 collected.

Ogawa et al. [2005] identify essential details of a method for accurately monitoring the course of charcoal carbon sequestration. In brief, the method recommends the monitoring of (i) the source, quantity and quality of biomass consumed in the carbonisation, (ii) carbonization method, charcoal yield, carbon composition in charcoal, external heating requirement, and (iii) the identification of non-fuel use of charcoal and related carbon stability aspects.

Further work on monitoring, inventories and accounting is necessary if negative emission biomass technologies shall be incorporated in the accounting systems of international climate regimes.

Global Potentials

Some modelling using global energy system models has been performed with consideration taken to the possibilities of BECS (see the introduction to this chapter for some

---

[36] "Fixed carbon" indicates carbon in biochar. Fixed carbon (%) = 100 % - (moisture % + ash % + volatile %). Recovery of fixed carbon indicates the percentage of (Fixed carbon)/(Carbon from fuelwood).

details and references). In relation to charcoal carbon sequestration, Lehman et al. [2005] estimated the total technical sequestration potential in the range 5.5 to 9.5 Pg C/year. Further modelling exercises including both BECS and charcoal sequestration technology are called for.

Finally, it is important to point out that there are still a number of issues open with respect to the environmental and social sustainability of large-scale global applications of negative emission biomass technologies which go beyond the scope of this paper. Given that land markets are rather sticky, large infrastructure investments are necessary and diffusion times will span several decades. In particular, it has been observed that large scale biomass plantations have faced political and social objection both in the developed and developing world. It can be concluded that as long as the high option value, the high environmental and potentially social values of negative emission biomass technologies are not recognized, it will fail to be developed adequately as a competitive mitigation and energy technology cluster. Furthermore, as already noted, technological spillovers can be expected from the implementation of CCS technologies with fossil fuels.

CONCLUSION

This chapter has presented a biomass energy with CO_2 capture and storage (BECS) implementation scenario study. Investments in BECS in a pulp and paper mill environment were analysed within a real options framework. Uncertainty was considered in the economic modelling through the use of stochastically correlated price processes of one input price (biomass) and two output prices (electricity and CO_2 emission permits) that are consistent with shadow price trajectories of a large-scale global energy model. The analysis suggests that (in the range of scales considered), BECS can be economically feasible within approximately 40 years. Combined with a number of economic factors, the uncertainties on the competitiveness of the negative emission biomass technology cluster is larger than that of more conventional mitigation technologies. Therefore we provided a discussion concerning Research and Development efforts that would allow increasing certainty and buying down costs.

ACKNOWLEDGEMENTS

The authors gratefully acknowledge financial support from the Kempe foundations, the Swedish Association of Graduate Engineers, the IIASA Forestry program and the EU project-Integrated Sink Enhancement Assessment (INSEA), contract number: SSPI-CT-2003/503614. The authors are also grateful to Andre Faaij and Yasuyuki Okimori for providing helpful comments on early drafts.

REFERENCES

Ådahl, A.; Harvey, S. In: *Process industry energy projects in a climate change conscious economy*; Ådahl, A; PhD thesis, Chalmers University of Technology, Goteborg, 2004.

AspenTech (2005). *http://www.aspentech.com*

Audus, H.; Freund, P. In *Proceedings 7th International Conference on Greenhouse Gas Control Technologies;* Rubin, E. S.; Keith, D. W.; Gilboy, C. F.; Eds.; IEA Greenhouse Gas Programme: Cheltenham, 2004.

Azar, Ch.; Lindgren, K.; Larson, E. D.; Möllersten, K. *Climatic Change.* 2006, 74, 47–79.

Baciocchi, R.; Storti, G.; Mazzotti, M. *Proceedings 8th Greenhouse Gas Control Technologies Conference*, Trondheim, 2006.

Berglin, N.; Eriksson, T.; Berntsson, T. *Proceedings 2nd Johan Gullichsen Colloquium*, 1999, 55-67.

Börjesson, P.; Gustavsson, L. *Energy.* 1996, 21, 747-764.

Bradshaw, J.; Bachu, S.; Bonijoly, D.; Burruss, R.; Holloway, S.; Christensen, N. P.; Mathiassen, O. M. *Proceedings 8th Greenhouse Gas Control Technologies Conference*, Trondheim, 2006.

Brandberg, L.; Ekbom, T.; Hjerpe, C.-J.; Hjortsberg, H.; Landälv, I.; Sävbark, B. *BioMeet: Planning of Biomass-Based Methanol Energy Combine* - Trollhättan Region; Report no XVII/4.1030/Z/98/368; Ecotraffic Research and Development AB and Nykomb Synergetics AB: Stockholm, 2000.

Browne T. *Pulp Pap.* 2003, 104, 27-31.

Bruce D. M. *Pulp Pap.* 2000, 101, 35-38.

Ceila, M.A; Bachu, S. In Gale, J.; Kaya, Y.; Eds.; *Proceedings 6th International Conference on Greenhouse Gas Control Technologies*, Elsevier Science: Oxford, 2003. .

Chladná, Z.; Chladný, M.; Möllersten, K.; Obersteiner, M. *Investment under Multiple Uncertainties: The Case of Future Pulp and Paper Mills*; IR-04-077; International Institute for Applied Systems Analysis (IIASA): Laxenburg, 2004.

David, J.; Herzog, H. J. In *Proceedings: 5th International Conference on Greenhouse Gas Control Technologies*; Williams, D. J.; Durie, R. A.; McMullan, P.; Paulson, C. A. J.; Smith, A. Y.; Eeds.; CSIRO Publishing: Collingwood, 2000, pp 985-990.

Day, D.; Evans, R. J.; Lee, J. W.; Reicosky, D. *Energy.* 2005, 30, 2558-2579.

Dixit, A.; Pindyck, R. S. *Investment under Uncertainty*; Princeton University Press: Princeton, NJ, 1994.

DOE (U.S. Department of Energy). *Carbon Sequestration, Research and Development.* U.S. Department of Energy:Washington, DC, 1999.

Dornburg, V.; Faaij, A. *Biomass and Bioenergy.* 2001, 21, 91-108.

Dowaki, K.; Ishitani, H.; Matsuhashi, R.; Sam, N. *Technology.* 2002, 8, 193-204.

Ekström, C.; Blumer, M.; Cavani, A.; Hedberg, M.; Hinderson, A.; Svensson, C. G.; Westermark, M.; Erlström, M.; Hagenfeldt, S. *Technologies and costs in Sweden for capture and storage of CO_2 from combustion of fossil fuels in production of power, heat, and/or transportation fuels*; Vattenfall Utveckling AB: Stockholm, 1997. (In Swedish).

FAO (United Nations Food and Agriculture Organization) (2005). *Statistical Databases.* http://www.fao.org.

Faaij, A.; Meuleman, B.; Van Ree R. Long term perspectives of biomass integrated gasification/combined cycle (BIG/CC) technology; performance and costs; *EWAB* 9840: NOVEM, 1998.

Falkowski, P.; Scholes, R.J.; Boyle, E.; Canadell, J.; Canfield, D.; Elser, J.; Gruber, N.; Hibbard, K.; Högberg, P.; Linder, S.; Mackenzie, F.T.; Moore, B.; Pedersen, T.; Rosenthal, Y.; Seitzinger, S.; Smetacek, V.; Steffen, W. *Science*. 2000, 290, 291-296.

Freund, P.; Davison, J. In *Proceedings IPPC workshop on carbon dioxide capture and storage*; Regina, 2002.

Greene, N.. *Growing energy: how biofuels can help end America's growing oil dependence*; National Resources Defence Council, New York, NY, 2004.

Grönkvist, S.; Sjödin, J.; Westermark, M. *Int. J. Energ. Res.* 2003, 27, 601-613.

Grönkvist, S.; Möllersten, K.; Pingoud, K. *Mitigation and adaptation strategies for global change*. Published online March 2006.

Ha-Duong, M.; Keith, D. W. *Clean Technologies and Environmental Policy*. 2003, 5, 181-189.

Hamelinck, C. N. *Outlook for advanced biofuels*; Ph.D. thesis. Utrecht University; Utrecht, 2004.

Hochenauer, C.; Hohenwarter, U.; Sanz, W.; Schlamadinger, B. *Proceedings ASME Turbo Expo*. 2004, Vienna, 2004.

IEA (International Energy Agency). *Transmission of CO_2 and energy*; Report no PH4/6; International Energy Agency Greenhouse Gas Research and Develoment Programme: Stoke Orchard, 2002.

IEA (International Energy Agency). *Carbon dioxide capture and storage issues – Accounting and baselines under the United Nations framework convention on climate change* (UNFCCC); International Energy Agency: Paris, 2004.

IPCC (Intergovernmental Panel on Climate Change). *Climate Change 2001: The scientific basis*; Cambridge University Press: Cambridge, 2001a.

IPCC (Intergovernmental Panel on Climate Change). *Climate Change 2001: Mitigation;* Cambridge University Press: Cambridge, 2001b.

IPCC (Intergovernmental Panel on Climate Change). *Special Report on Carbon Capture and Storage;* Cambridge University Press: Cambridge, United Kingdom and New York, NY, USA, 2005.

Karlsson, Å. *Comparative assessment of fuel-based systems for space heating*. Doctoral thesis. Lund, Sweden: Lund University, Department of Technology and Society, 2003.

Keith, D. W. *Climatic Change*. 2001, 49, 1-10.

Keith D. W.; Rhodes J. S. *Climatic Change*. 2002, 54, 375-377.

Kumar, A.; Cameron, J. B.; Flynn, P. C. *Biomass Bioenergy*. 2002, 24, 445-464.

Lackner, K. S. *Science*. 2003, 300, 1677-1678.

Larson, E. D.; Marrison, C. I. *J. Eng. Gas Turbines Power*. 1997, 119, 285–290.

Larson, E. D.; Kreutz, T. G; Consonni, S. *J. Eng. Gas Turbines Power*. 1999,121, 394-400.

Larson, E.; Consonni, S.; Kreutz, T. *J. Eng. Gas Turbines Power*. 2000, 122, 255-261.

Larson, E. D.; Jin, H.; Celik, F. E. *Gasification-based fuels and electricity production from biomass, without and with carbon capture and storage*. Princeton, NJ: PEI, Princeton Univ; 2005.

Lehmann, J.; Gaunt, J.; Rondon M. *Mitigation and Adaptation Strategies for Global Change* (forthcoming).

Lindblom, M. Chemrec AB. *Personal communication.* 2001.

Martin, N.; Anglani, N.; Einstein, D.; Khrushch, M.; Worrell, E.; Price L. K. *Opportunities to improve energy efficiency and reduce greenhouse gas emissions in the US pulp and paper industry*; Report no. LBNL-46141; Lawrence Berkeley National Laboratory: Berkeley, CA, 2000.

Maunsbach, K.; Isaksson, A.; Yan, J.; Svedberg, G.; Eidensten, L. J. *J. Eng. Gas Turbines Power.* 2001, 123, 734-740.

Messner, S.; Schrattenholzer, L. *Energy.* 2000, 25, 267-282.

Miner, R.; Lucier, A. *A Value Chain Assessment of Climate Change and Energy Issues Affecting the Global Forest-Based Industry; National Council for Air and Stream Improvement* (NCASI): Research Triangle Park, NC, 2004.

Möllersten, K.; Yan, J. *World Resour. Rev.* 2001, 13, 509-525.

Möllersten, K. *Opportunities for CO_2 reductions and CO_2-lean energy systems in pulp and paper mills*; TRITA-KET R161; Royal Institute of Technology: Stockholm, 2002.

Möllersten, K.; Yan, J.; Moreira, J.R. *Biomass Bioenergy.* 2003a, 25, 273-285.

Möllersten, K.; Gao, L.; Yan, J. *Mitigation and Adaptations Strategies for Global Change.* Published online August 2006.

Mun, J. *Real Options Analysis: Tools and Techniques for Valuing Strategic Investments and Decisions.* Wiley finance series, USA, 2002.

Nakicenovic, N.; Riahi, K. *An Assessment of Technological Change across Selected Energy Scenarios;* RR-02-005 (Reprint from a WEC report); International Institute for Applied Systems Analysis: Laxenburg, 2002.

National Research Council. *Abrupt Climate Change – Inevitable Surprises.* National Academy Press, Washington, DC, 2002.

Nguyen, M. H.; Prince, R. G. H. *Biomass Bioenergy.* 1996, 10, 361-365.

Obersteiner, M.; Azar, Ch.; Kauppi, P.; Möllersten, K.; Moreira, J.; Nilsson, S.; Read, P.; Riahi, K.; Schlamadinger, B.; Yamagata, Y.; Yan, J. van Ypersele J-P. *Science.* 2001, 294, 786-787.

Obersteiner, M.; Möllersten, K.; Azar, Ch.; Bohlin, B. *Climate risk management – Are we ignoring the obvious*? EuroScience Open Forum, Stockholm, 2004.

Ogawa, M.; Okimori, Y.; Takahashi F. *Mitigation and adaptation strategies for global change.* 2005 (forthcoming).

Ogden, J. M. In Gale, J.; Kaya, Y.; Eds.; *Proceedings 6th International Conference on Greenhouse Gas Control Technologies*, Elsevier Science: Oxford, 2003..

Okimori, Y.; Ogawa, M.; Takahashi, F. *Mitigation and adaptation strategies for global change.* 2003, 8, 261-280.

Rao, S.; Riahi, K *The Energy Journal* (in press).

Rhodes, J.; Keith, D. *Biomass Bioenergy,* 2005, 29, 440-450.

Riahi, K.; Roehrl, R. A. *Technol. Forecast. Soc. Change.* 2000, 63, 175-205.

Riahi, K.; Rubin, E. S.; Schrattenholzer, L. In Gale, J.; Kaya, Y.; Eds.; *Proceedings 6th International Conference on Greenhouse Gas Control Technologies*, Elsevier Science: Oxford, 2003. pp. 1095-1100.

Schlamadinger, B.; Apps, M.; Bohlin, F.; Gustavsson, L.; Jungmeier, G.; Marland, G.; Pingoud, K.; Savolainen, I. *Biomass Bioenergy.* 1997, 13, 359-375.

Seifritz, W. Int. J. *Hydrogen Energy.* 1993, 18, 405-407.

Siro, M. *First Finnish Kraft pulp and paper mill independent of oil as an energy source.* *Proceedings*: 1984 TAPPI Pulping Conference, 1984, 591-595.

Smith, S. J.; Brenkert, A.; Edmonds, J. *Proceedings 8th Greenhouse Gas Control Technologies Conference*, Trondheim, 2006.

STFI (Swedish Pulp and Paper Research Institute). *The Ecocyclic Pulp Mill, phase 1. Final report KAM1 1996-1999*; Swedish Pulp and Paper Research Institute: Stockholm, 2000.

Swedish Pulp and Paper Research Institute (STFI). *Ecocyclic Pulp Mill – "KAM", Final report 1996-2002*; KAM A100; Swedish Pulp and Paper Research Institute: Stockholm 2003.

Stigsson, L.; Berglin, N. In *Proceedings 6th International Conference on New Available Technologies at SPCI99*; Stockholm, 1999.

Strømman, A. Norweigan University of Science and Technology, Trondheim, Norway. *Personal communication*, 2003.

Svensson, R.; Odenberger, M.; Johnsson, F.; Strömberg, L. *Energy Convers. Manag. 2004,* 45, 2343-2353.

Thomas, C. D.; Cameron, A.; Green, R. E.; Bakkenes, M.; Beaumont, L. J.; Collingham, Y. C.; Erasmus, B. F. N.; Ferreira De Siqueira, M.; Grainger, A.; Hannah, L.; Huges, L.; Huntley, B.; Van Jaarsveld, A. S.; Midgley, G. F.; Miles, L.; Ortega-Huerta, M. A.; Townsend Peterson, A.; Phillips, O. L.; Williams, S. E. *Nature.* 2004, 427, 145-148.

UNFCCC (United Nations Framework Convention on Climate Change) (1992). *http://www.unfccc.int.*

Upton, B. H.; Mannisto, H. *Technologies for reducing carbon dioxide emissions: A resource manual for pulp, paper, and wood products manufacturers*; Report no. 01-05. National Council for Air and Stream Improvement: Research Triangle Park, NC, 2001.

Van Vuuren, D.; den Elzen, M.; Lucas, P.; Eickhout, B.; Strengers, B.; van Ruijven, B.; Wonink, S; van Houdt, R. *Climatic Change (*in press).

Veijonen, K., Vainikka, P. Järvinen, T., Alakangas, E. Biomass co-firing – an efficient way to reduce greenhouse gas emissions; *VTT Processes: Jyväskylä,* 2003.

Walsh, J. H. *Energy Convers. Manag.* 1993, 34, 1031.

Warnqvist, B. *Techno-economic evaluation of black liquor gasification processes*; Report no 698; Värmeforsk: Stockholm, 2000 (in Swedish).

Wigley, T. R.; Richels, R.; Edmonds, *J. Nature.* 1996, 359, 240-243.

Williams, R. H. *In Eco-Restructuring: Implications for Sustainable Development*; Ayres, R. Ed; United Nations University Press: Tokyo, 1998, 180–222.

Williams, R. H. In *Proceedings VIII Biennial Asilomar Conference on Transportation, Energy, and Environmental Policy*: Managing Transitions; Kurani, K. S.; Sperling D.; Eds., Transportation Research Board: Washington, DC, 2002.

In: Progress in Biomass and Bioenergy Research
Editor: Steven F. Warnmer, pp. 101-140

ISBN: 1-60021-328-6
© 2007 Nova Science Publishers, Inc.

Chapter 4

A REVIEW OF THE SOCIO-ECONOMIC AND ENVIRONMENTAL BENEFITS OF BIOMASS GASIFICATION BASED POWER PLANT: LESSONS FROM INDIA

Kakali Mukhopadhyay[*]

UNEP-NISD; Centre for Development and Environment Policy, Indian Institute of Management Calcutta, India; Joka, D.H.Road, Calcutta-700104, INDIA

ABSTRACT

There is a steady and continuing interest in biomass gasification in both the developed countries and developing countries. While the advanced countries are interested primarily from considerations of reduced emissions and waste utilisation, the developing countries look at biomass gasification as a means to augment commercial energy like electricity, diesel, fuel oil etc.

India, a tropical country with a vast geographical area is richly endowed with renewable energy sources like solar, wind, biomass which can play a crucial role in meeting end use energy needs in a decentralised manner. One of the major goals of the ninth and tenth five year plan is strengthening of infrastructure (energy, transport, communication, irrigation) in order to support the growth process on a sustainable basis. It is usually the tendency of the developing countries to equate development with economic growth and to further equate economic growth with energy consumption especially electricity. India being a developing country has also given due emphasis on strengthening its energy position accordingly. Moreover threat from Green House Gasses (GHG) also has caused worldwide concern. In India electric power generation is the largest source of GHG emissions. It accounts for 48% of carbon emitted. These concerns

[*] Consultant UNEP-NISD; Centre for Development and Environment Policy, Indian Institute of Management Calcutta, India; Joka, D.H.Road, Calcutta-700104, INDIA; kakali@iimcal.ac.in; kakali_mukhopadhyay@yahoo.co.in; Tel: 913324678300-04,; Fax: 913324678062

point towards more rational energy use strategies. The renewable and recycling process makes biomass possible to generate power without adding to air emissions.

Biomass (firewood, agricultural residue, and dung) is one of the main fuels in India, particularly in the energy-starved rural sector. The biomass power potential in India was 16,000 MW (excluding co-generation), but the achievement in this respect is negligible (Installed capacity - 630 MW Project under implementation - 630 MW, as on March 2005). It brings out the fact that much of the potential of biomass gasification is still unexplored. Globally, India is in the fourth position in generating power through biomass and with a huge potential, is poised to become a world leader in utilization of biomass.

According to the Planning Commission of India, in its Tenth Five Year Plan, announced that 26.10 per cent of the Indian populations are below the poverty line and mostly belongs to rural areas. The inequitable distribution has been evident from the fact that although 70% of India's population lives in the rural areas, only 29% of rural households have electricity supply as against 92% of urban households. Of the half a million or so villages in India, about 3, 10,000 villages have been declared to be electrified and 80,000 more villages remain completely un-electrified. There are a number of constraints to supply power to remote rural area such as small human settlements, geographically dispersed villages, seasonally of loads etc. In the absence of adequate network and hence supply of power to remote rural areas the household depend largely on primary energy sources like kerosene and diesel for lighting. No commercial investments in micro enterprises can therefore be made by either individuals or companies without installing diesel generators which have a very high generating cost. Biomass gasifier is a leading option in that respect. Besides, the supply of power to remote rural areas from the centralised grid is not competitive than a modern biomass gasification based decentralised power plant. Estimate from an Indian village shows that modest 50 kW of installed capacity per village will lead to total saving of 52000 million Rs (Rs 5200 Crore / 1100 million US $) in power plant investments. In energy terms, the saving in TandD losses will release a generation capacity of 800 MW for profitable sale. Reduced pollution and reduction of CO_2 emissions will be the other advantages of a decentralised renewable energy based system for the rural areas.

The purpose of the present paper is to evaluate the rural electrification programme in India undertaken by the Ministry of Non Conventional Energy Sources (MNES), Government of India, through biomass gasifier power plant. It explores the eradication of poverty that has been made possible by introducing biomass gasification based power plant in remote rural areas in India. Creation of jobs in the power stations, small-scale business, commerce and industries and also improvement in the quality of life is assessed. The paper concludes with policy options relevance for the other developing countries.

1. INTRODUCTION

Electricity is the key factor to the economic and social development of any country. During the last five decades, the demand for electricity has increased manifold in India, primarily due to the rapid rate of urbanization and industrialization. In India demand for electricity is growing at 7% annually, particularly in rural areas. However, its annual per capita consumption is less than 400 kWh, as compared to a world average of over 2,000 kWh (MNES, 2002).

Over 70% of the population of India live in villages. It is a matter of shame for all of us even 58 years after Independence, 63 per cent of all rural households in India do not have electricity and use kerosene for lighting. Even for those rural areas, which are electrified, there is a tremendous shortage of power supply. Thus it is not uncommon for these areas to have 10-15 hours of blackouts and brownouts every day. There is a shortfall of about 15,000-20,000 mw of electricity in the country and we require about 140,000 mw of additional capacity by 2010 with an estimated outlay of Rs 5, 50,000 crores (MNES, 2004). Moreover, with any problems in the national grid, rural areas are mostly affected because the state electricity boards provide urban areas with electricity on priority basis. Further, the electricity supply to low-load rural areas is characterized by high transmission and distribution costs and losses and subsidized pricing. So the power generation and supply situation is grim, with shortages in installed capacity and peak power supply. Because of acute shortage of electricity, industrial growth and general life in the country is also affected seriously. Educational facilities for higher studies do not exist in these villages and sophisticated hospitals or industries are also absent due to lack of electricity supply.

Furthermore, most of the villages in India is situated in remote rural areas and hence cannot be connected to the normal conventional grid. This is a vital problem. Thus, the sole option is to consider non-conventional sources of energy instead of considering fossil fuel based electricity.

India, a tropical country with a vast geographical area is richly endowed with renewable energy sources like solar, wind, biomass which can play a crucial role in meeting end use energy needs in a decentralised manner. India's need for power is growing at a remarkable rate and this requirement is being met by both commercial and renewable energy sources. India, today, has a total installed capacity of about 3400 MW of power from renewables, which is over 3% of the total power generation capacity in the country, still leaving a large capacity untapped. Annual electricity generation and consumption have nearly doubled since 1990. Electricity generation has grown from 275.5 billion kilowatt-hours (kWh) to 547.2 kWh, while consumption has grown from 257.1 kWh to 510.1 kWh. The country's projected increase in electricity consumption, of between 2.6 per cent and 4.5 per cent up to 2020, is the highest for any major country (UNDP, 2005).

An examination of India's primary energy balance shows that renewable account for about 33% of the primary energy consumption in India. Of this, the major contributor is traditional bio-mass that is used for electricity generation from gasifier plants.

Thus renewable energy development programme is gaining momentum in India. It has emerged as a viable option to achieve the goal of sustainable development (Thomas, 2002). As we know that coal is the major feeding item for the conventional electricity generation. Coal-based thermal power accounts for 70% of the installed capacity. (MNES report, 2004). Coal-based power generation is characterized by local and regional environmental degradation as well as greenhouse gas emissions, leading to climate change. Coal mining, storage, transportation, and combustion lead to environmental degradation. Coal combustion for power generation is the dominant source of greenhouse gases, both in India and globally. With global concern over greenhouse warming, India, like other developed and developing countries needs to consider this aspect as one of its guiding principles in selecting energy options in the medium and long term. Coal combustion also leads to the emission of oxides of sulphur and nitrogen, and production of large amounts of ash. This results in environmental pollution for the population residing around the power plant as well as further a field. Currently more crucial for developing countries are local air pollution, land degradation, and waste disposal problems arising from coal use. Thus, the search is on for an environmentally sound alternative for meeting the power needs, particularly of rural areas in developing countries. Adverse local environment impacts (SOx, NOx, SPM) and global environmental impacts (green house gas emissions mainly due to carbon dioxide) associated with fossil fuel use have resulted in an increased emphasis on renewables. India has now the world's largest programme for deployment of renewable energy products and systems (Thomas, 2002). The spread of various renewable energy technologies in the country has been supported by variety of incentives and policy measures.

Table 1.1 shows a listing of some of the commonly used renewable options. Renewables can be used for ace heating, cooling, water pumping, cooking and for almost any endues that is presently met by fossil fuels.

Table 1.1. Renewable Energy Options

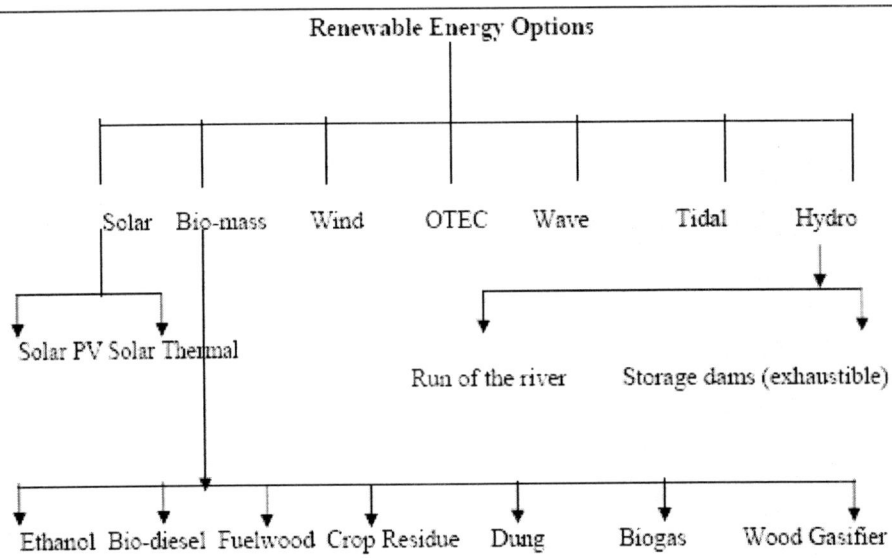

*Modern Renewables
Source: Planning Commission, 2005

Several studies, surveys and documents show that biomass is the most convenient option for decentralized power generation in rural India compared to other renewables. The arguments are discussed below.

Why Biomass

India being an agrarian economy there is easy availability of agricultural based mass which can be used to generate energy. The term 'biomass' refers to organic matter, which can be converted to energy. Biomass is available all round the year. It is cheap, widely available, easy to transport, store, and has no environmental hazards. It can be obtained from plantation of land having no competitive use. Biomass-based power generation systems, linked to plantations on wasteland, simultaneously address the vital issues of wastelands development, environmental restoration, rural employment generation, and generation of power with no distribution losses. It can be combined with production of other useful products, making it an attractive by product. Some of the most common biomass fuels are wood, agricultural residues, and crops grown specifically for energy. In addition, it is possible to convert municipal waste, manure or agricultural products into valuable fuels for transportation, industry, and even residential use.

Burning this biomass is the easiest and oldest method of generating energy. Among the various renewable energy sources, biomass conversion technologies appear to be one of the best suited for conversion to shaft power/electricity through gasifier. The gasifier is essentially a chemical reactor, where several thermo-chemical processes such as pyrolysis, combustion and reduction of biomass take place under controlled conditions. (Rehman, 2002). India has done some pioneering work in this area for many years and built open-top gasifiers and integrated them with internal combustion engines or boilers.

Biomass exists in rural areas and needs to be tapped to provide not only electricity but also water to irrigate and cultivate fields to further increase production of biomass (either as a main product or as a by-product). The various applications of biomass energy include thermal or heat, mechanical water pumping for irrigation and power generation including village electrification. The availability of waste biomass from the biomass gasifier plant to be used as fertilizer is an added advantage.

As a renewable fuel, biomass is used in nearly every corner of the developing world as a source of heat, particularly in the domestic sector. Unlike other renewables, biomass is a versatile source of energy, particularly attractive for decentralized applications which can be converted to 'modern' forms such as liquid and gaseous fuels, electricity, and process heat. Bioenergy also permits operation at varying scales. For example, small-scale (5–10 kW), medium-scale (1–10 MW) and large-sale (about 50 MW) electricity generation systems or biogas plants of a few cubic metres (Indian and Chinese family plants for cooking) to several thousand cubic metres (Danish systems for heat and electricity). This variety of scales is useful for power generation for decentralized applications at the village level as well as for supply to the national grids. Unlike wind, solar or micro-hydroelectric systems, modern biomass energy systems could be set up in virtually any location where plants can be grown. (WEC, 1993)

Renewables such as solar, wind, and micro-hydro require 'spare' or additional capacity to produce adequate energy when the conditions are favourable, such as water flow or wind speed. This intermittent feature of such renewable energy sources necessitates electricity storage facilities, especially with small and local systems, if sustained demand requirements are to be met (WEC1993). Bioenergy sources such as producer gas systems do not require electricity storage. This is an important advantage.

Currently, biomass contributes 15 per cent of the total energy supply worldwide and 40 per cent of this energy is consumed in developing countries, mostly in the rural and traditional sectors of the economy.

According to WEC (1993), under the minimum case scenario, modern biomass is the most important of the renewables and is projected to account for 45 per cent of the new contribution by renewables to world energy by 2020. In the maximum scenario projections, modern biomass will account for 42 per cent of the total renewable energy contribution by 2020. The US Department of Energy suggests that biomass power will be the most important renewable energy option for the next quarter of a century and has projected a possible 25 GW in installed power in the US alone by 2010 (USDOE 1993). This clearly demonstrates a dominant role for biomass energy among all the renewable.

Most common source of biomass is wood waste and agricultural wastes. In India development of biomass gasification has received serious attention with establishment of biomass research centers and gasifier action research centers at various locations spread all over the country. These institutions have played a key role in upgradation and adoption of suitable technologies, testing, monitoring and development of biomass gasification systems. Studies reveal that the low grade of land suitable only for scrub vegetation can be turned to advantage and form an excellent source of biomass – fast growing trees and shrubs. India has actively promoted research and development programmes for efficient utilization of biomass and agro wastes and further efforts are on.

In short, biomass energy systems offer an opportunity for sustainable (as biomass can be grown sustainable), self-reliant (biomass is available in all countries and can be converted to gaseous or liquid fuels and electricity, leading to a decrease in imports of oil), and equitable development (between and within countries due to universal availability of biomass and the fact that decentralized bio-energy systems lead to local control and employment).

As far as biomass-based power generation is concerned, the commissioned biomass power capacity reached 290 MW (52 projects) and the commissioned co-generation capacity installed mainly at sugar mills reached 437 MW (57 projects) by the end of year 2004. In the area of small-scale biomass gasification, significant developments in technology have made India a world leader. A total capacity of 55 MW (1817 projects) of biomass gasifier system has so far been installed in India, mainly for stand-alone applications. Biomass gasifiers capable of producing power from a few kilowatts up to 500 kW have been successfully developed indigenously and are also now being exported to the developing countries of Asia and Latin America, and also Europe and USA (MNES, 2005).

In India more than 2000 gasifiers are estimated to have been established with a capacity in excess of 22 MW and a number of villages have been electrified with biomass gasifier based generators. These gasifiers mainly consumed, fuel wood (200-300 million tonnes), animal waste (80- 100 million tonnes) and crop residues (100-120 million tonnes) annually as the main biomass fuels. Fuel wood contributes nearly 60 per cent of the total biomass energy in India estimated between 200-300 million tonnes. There are immense scope and potential to be acquired from biomass gasification i.e., water pumping, electricity generation - 3 to 1 MW power plants, heat generation: for cooking gas – smokeless environment and rural electrification ensuring better healthcare, better education and improved quality of life (MNES, 2005).

There is no doubt that the environment friendly electricity generation is only possible through renewables. Since the central focus of this paper is on biomass gasification and corresponding generation, its eco- friendly role also receives proper attention. When biomass is used to produce power, the carbon dioxide released at the power plant is recycled back into the re-growth of new biomass. This renewable and recycling process makes it possible to generate power without adding to air emissions. Due to the non-availability of the sufficient resources and a considerable amount of emission of pollutants from commercial energy, it is now being felt that renewable energy has to be utilised to a greater extent. In India electric power generation is the largest source of GHG emissions and accounts for 48% of carbon emitted. These concerns point towards more rational energy use strategies.

The statistics shows that estimated biomass power potential is always high but its installed capacity is low, though its commercial viability is quite well known. This fact is quite apparent in tables 1.2 and 1.3. Biomass collection logistics and competing uses of biomass are the most serious constraints in the widespread utilization of this resource for power generation. For the grid interactive (table 1.2) case the total cumulative installed capacity is 7097.54 MW as on 31December 2005 but the share of biomass power is nominal. As against the 16,000MW estimated potential of biomass power, a cumulative installed capacity of 376 MW has been commissioned till 31.12.05. Maximum capacity is under implementation in Tamil Nadu, Andhra Pradesh, Karnataka, UP, Maharashtra, Punjab and Haryana. The cumulative installed capacity is further low for decentralized energy systems (table 1.3).

Table 1.2. Grid-interactive renewable power

Source/ system	Estimated potential(MW)	Cumulative installed capacity(as on 31 December 2005) (MW)
Wind power	45 000	4434.00
Biomass power	16 000	376.00
Bagasse cogeneration	3500	491.00
Small hydro (up to 25 MW)	15 000	1747.98
Waste-to-energy	2700	45.76
Solar photovoltaic	20 MW per km2	2.80
	Total	**7097.54**

Source: Planning Commission, 2005

Table 1.3. Decentralized renewable energy systems

Source/system Estimated potential Cumulative installed capacity (as on 31 December 2005)
1. Family-size biogas plants 120 lakh 38.00 lakh
2. Community/institutional/night soil biogas plant — 3952 nos
3. Improved chulha 12 crore 3.52 crore
4. Solar photovoltaic systems 20 MW/km2
5. Solar water heating systems 140 million m2 collector area 1.5 million m2 collector area
7. Solar PV pumps — 6818 nos
8. Wind pumps — 1087 nos
9. Hybrid systems — 410 kW
10. Biomass gasifiers — 69 MW

Source: Planning Commission, 2005

Biomass for power generation has been recognized as an important component of the renewable energy programme in India and this is reflected in the priority by the MNES.

Recently, the Prime Minister has set up the Rural Electricity Supply Technology (REST) mission in the Union ministry of power. The State Governments have been directed to take up the electrification of 62,000 villages through the Electricity Boards under the traditional rural electrification programmes by 2007 under the Pradhan Mantri Gramodhaya Yojna.. The Government of India has also directed MNES to take up renewable energy based electrification of 18,000 villages in remote and inaccessible parts of the country by 2012. These villages would be electrified through decentralised plants based on biomass, gasification of biomass, hydel power, solar thermal power etc. The action plans so far implemented or to be undertaken has been portrayed in section 2.

The above background simply states that Biomass potential is enormous in India and it can be considered as a feasible option in all respect among other renewable. Though the potential is high in comparison to other renewable, its application is not up to the desired level, however.

The **purpose of the paper** is to evaluate the feasibility of the biomass gasifier based power plants for rural electrification. The feasibility mentioned here basically encompasses the benefits acquired by the villagers due to the power plant. The evaluation will also help the policy makers and researchers to get a comprehensive overview of rural electrification in India through decentralized power plant. Further this will encourage those unelectrified villages going to undertake biomass gasifier power plant as an option.

The rest of the paper is structured as follows. Section 2 reviews implementation of the rural electrification programme, actions and strategies formulated by the Government of India and the various programmes undertaken by the Ministry of Power. Section 3 describes the village level case studies mainly in respect of rural electrification. It evaluates the socio-economic and environmental benefits of the installed BGBPP. Overall assessment of the case studies are explained in section 4 and section 5 concludes the paper with policy options.

2. RURAL ELECTRIFICATION PROGRAMME THROUGH BIOMASS GASIFICATION BASED POWER PLANT

Rural electrification programme is a common agenda in every five year plans in India. Its acute necessity has also been reflected in various plans. But the proper implementation of the programme is a fundamental problem. In this section the paper presents the various schemes towards rural electrification undertaken by MNES, Government of India so far.

Rural Electrification Initiatives

Rural electrification programmes began in the 1950s as a social amenity, but they gathered importance in the mid-sixties as a source of energy for pumping irrigation water. The Rural Electrification Corporation was formed in 1969 with the task of electrifying all villages in India. Significant success in this effort has been achieved and 84 per cent of the villages were electrified by October 1993 (CMIE 1994). Many major states, such as Andhra Pradesh, Gujarat, Karnataka, Maharashtra, and Punjab, have the distinction of electrifying nearly 100 per cent of villages. However, in West Bengal on 73 per cent of the villages are electrified, while the lowest level of electrification is in Bihar where only 70 per cent of villages are electrified. Thus the achievement of the rural electrification programme has been excellent; for example between 1980 and 1991 the number of villages newly electrified was 212634. However, there are many critics of the rural electrification programme, who argue that there are many villages (no data are available) where electricity transmission lines pass through the village and yet none of the households or farms has benefited from electrification. Such villages are, however, classified as 'electrified'. There could be some truth in these misgivings since the data from some states, such as Bihar and Uttar Pradesh show that only 4.4 and 9.3 per cent, respectively, or rural homes are actually electrified (NSS 1992). What is important is that at least 70 000 villages are yet to be connected to the grid, and even more important is that these villages could be remote, requiring very large investments to connect them to the grid. This assumption is based on the logic that the state electricity utilities, wanting to show progress and achieve physical targets, start with villages near existing grid transmission lines and leave all the remote and difficult villages to the last. These villages provide the first opportunity of alternative power generation systems.

During the year 2002-2003, 3056 inhabited villages were electrified as on 31.12.2002 and 213618 pumpsets /tubewells energized as on 30.11.2002. Cumulatively 509678 villages have been electrified and 13355909 electric irrigation pumpsets have been energized as on 30.11.2002(Ministry of Power, 2003). Table 2.1 shows the status of rural electrification recently. The major percentage of unelectrified villages and corresponding household is concentrated in Jharkhand, Bihar and Uttar Pradesh. The number of remote villages in India is large mostly concentrated in the eastern part of Indian peninsula. The number of the remote village electrification is also growing (table 2.2). Towards this end it is better to check the state wise total number of biomass gasifier installation (table2. 3).

A total capacity of 55.105 MW has so far been installed, mainly for stand-alone applications. The most worth mentioning plants in this respect are: 1) A 5 x 100 KW biomass gasifier installation on Gosaba Island in Sunderbans area of West Bengal is being successfully run on a commercial basis to provide electricity to the inhabitants of the Island through a local grid. 2) A 4X250 kW (1.00 MW) Biomass Gasifier based project has alsdo been commissioned at Khtrichera, Tripura for village electrification.

Table 2.1. Status of Rural Electrification - Selected States

State	Villages to be electrified	%village un-electrified	Households to be electrified	%households un-electrified
Jharkhand	22,920	78%	3,422,425	90%
Bihar	20,449	53%	12,010,504	95%
Uttar Pradesh	40,389	42%	15,505,786	80%
Assam	5,640	23%	3,522,331	84%
Orissa	9,682	21%	6,651,135	81%
West Bengal	7,694	20%	8,899,353	80%

Source: Ministry of Power, 2003

The purpose of the present study is to evaluate the rural electrification programme in India undertaken by the Ministry of Non Conventional Energy Sources (MNES), Government of India, through biomass gasifier power plant. Recently, MNES provides power to 15 villages in Karnataka by the utilization of biomass. The project aims at sustainable transformation of energy in rural areas of Karnataka, it is providing power to 15 villages in the Tumkur district of the state. The energy is produced through biomass gasifiers for stand-alone applications and supplying to the villages. The stand-alone gasifiers are located in every village within the capacity of 60 to 100 kW (The Business Standard, 25 February 2000).

Table 2.2. Remote village electrification

	Remote villages/hamlets electrified through RE
Item	(as on 31 December 2005)
Remote village	2195 remote villages
Electrification	594 remote hamlet

Source: MNES, 2005

Table 2.3. State-wise Cumulative Total Number and Capacity of Biomass Gasifiers Installed upto 30th June, 2003

S.No.	State	Cumulative	
		No. of systems	Capacity (in kW)
1.	Andhra Pradesh	231	15384
2.	Arunachal Pradesh	3	180
3.	Assam	6	123
4.	Bihar	2	20
5.	Chhatisgarh	1	500
6.	Goa	3	22
7.	Gujarat	237	11961
8.	Haryana	25	964
9.	Himachal Pradesh	2	7
10.	J and K	4	120
11.	Karnataka	476	4499
12.	Kerala	13	725
13.	Madhya Pradesh	144	4529
14.	Maharashtra	316	3823
15.	Mizoram	2	200
16.	Orissa	16	72
17.	Punjab	27	700
18.	Rajasthan	21	218
19.	Tamilnadu	83	2653
20.	Tripura	4	1000
21.	Uttar Pradesh	50	2746
22.	West Bengal	27	4100
23.	AandN Island	17	167
24.	Delhi	16	74
25.	Others	91	318
	Total	**1817**	**55105**

Source: MNES, 2004

In pursuance of the national agriculture policy, a National Biomass Resource Atlas is being prepared with the specific intention of boosting power generation from biomass. The national agriculture policy had called for increasing power generation from renewable sources for meeting the needs of agriculture. The National Biomass Resource Assessment Programme has been assigned this task. 150 taluka level studies have been completed and another 175 are being taken up (The Financial Express, 4 September 2000). According to a recent initial assessment made by the MNES about 500 million tonnes of biomass is generated every year from crop residues, bagasse, agro residues, and forest sources. The newspaper also says that out of this, only 170 million tonnes is used every year for power generation.

The action and strategies for rural electrification taken by the Government of India through Ministry of Power and MNES under various planning period is explained below:

Rajiv Gandhi Gram Vidyutikaran Yojana (RGGVY) ---- This initiative is to provide electricity access to all households in five years covering the entire country. This programme provides for ninety per cent capital subsidy for rural electrification projects covering to electrify the 1, 25,000 unelectrified villages. It will connect all the estimated 2.34 crore unelectrified households below the poverty line (BPL) with 90 percent subsidy on connecting costs and augment the backbone network in all the already electrified 4.62 lakh house holds by 2010. The 5.46 crore households above the poverty line, which are currently unelectrified, are expected to get electricity connection on their own without any subsidy.

Such expansion of connectivity will require a corresponding expansion of supply capability. Given the present widespread and acute shortage of power in many states, special action is needed to facilitate and encourage decentralized distributed generation (DG) system so that community can take their destiny in their own hands instead of waiting for utility companies to supply electricity reliably.

As per the NSS 55th Round Survey in 1999-2000, among the households in rural areas who had electricity, the households who belonged to the poorest 5 percent of all rural households, spent more than Rs.300 per year for electricity. Thus a charge of Rs.1.0 per kWhr for the first 30 units per month should be within the capacity and willingness of even the poorest 5 percent households. Above 30 units per month the normal charge should be levied. For this electricity would have to be metered. An effective way of targeting the subsidy to BPL households will have to be found.

Increasing the level of rural household electrification from 44 per cent (2001) to the targeted 100 per cent by 2012 would lead to a huge growth in both demand and consumption. The Rs. 16,000 crores (nearly US$ 3.5 billion) outlay for the scheme, also opens up big opportunities for electrical equipment manufacturers.

Pradhan Mantri Gramodaya Yojana (PMGY)

Rural Electrification was included under Pradhan Mantri Gramodaya Yojana (PMGY) from 2001-02 to achieve human development at the village level. The six components of PMGY now are: Primary Health, Primary Education, Rural Drinking Water, Rural Shelter, Nutrition and Rural Electrification.

During 2002-03, the PMGY is being administered by the Planning Commission. Under the revised guidelines, the States would have flexibility to decide their inter-se allocation of ACA among the six PMGY sectors as per their own plan priorities and discretion. The funds for village electrification are available as Additional Central Assistance with 90% grant and 10% loan for the special category States, and 30% grant and 70% loan for other States. Government has released Rs.36066.35 lakhs to various States as first installment (50%) under PMGY for 2002-03. The Centre proposes to electrify 62,000 villages through grid power, during the 10th Five-Year Plan (2002-07) under the Pradhan Mantri Gramodhaya Yojna. The Centre hopes to achieve 100 per cent village electrification in the current Plan. Another 18,000 remote villages will be electrified through non-conventional energy as grid power could prove uneconomical. These villages would be electrified through decentralised plants based on biomass, gasification of biomass, hydel power, solar thermal power etc.

Minimum Needs Programme (MNP)

The revised criteria for the MNP components of rural electrification adopted since the beginning of 7th Plan are as under:

(a) all North-Eastern hilly States;

(b) all States with less than 65% electrification and in these States those districts will be taken up which has less than 65% electrification provided that districts having least percentage coverage will be given priority over the others; and

(c) all areas including in the tribal sub plan. During 2002-03 Rs.600 crores have been allocated to the eligible States under MNP. The break up is follows.

Table 2.4. Minimum Needs Programme allocated outlay (Rs. in lakhs)

S.No.	States	Amount
1.	Arunachal Pradesh	1200
2.	Assam	6000
3.	Bihar	6800
4.	Chattisgarh	800
5.	Himachal Pradesh	200
6.	Jharkhand	6800
7.	Madhya Pradesh	800
8.	Manipur	270
9.	Meghalaya	3000
10.	Nagaland	130
11.	Orissa	6000
12.	Uttar Pradesh	15000
13.	Uttaranchal	7000
14.	West Bengal	6000
	Total	60000

Source: Ministry of Power, 2003

Rural Electricity Supply Technology Mission (REST)

Distributed Generation has been identified as one of the technologies for ensuring supply of power in rural areas by way of setting up of small generating units based on a variety of local funds along with localized distribution. The electricity distribution in the rural areas is characterized by low density, high cost of delivery, poor availability of supply of power and commercially unviable on account of high faxed cost and high variable cost.

In order to utilize technology in proving for an affordable solution in making available electricity in rural areas, it has been decided to constitute the Rural Electricity Supply Technology Mission (REST) under the auspices of Minister of Power. The Mission would evolve a strategy based on technology which could provide for low cost power generation and low cost of delivery in the rural areas which can be managed by local institutions like Village Panchayats of Non-Government Organizations and to identify feasible size of generating units for different fuels, which are locally available and for mini and micro hydel projects. It is hoped through this mission to electrify all villages by 2010. Of the half a million or so villages in India, about 3, 10,000 villages have been declared to have been already electrified. According to government statistics, 80,000 more villages remain to be electrified. The State Governments have been directed to take up the electrification of 62,000 villages through the Electricity Boards under the traditional rural electrification programmes by 2007. The Government of India has also directed MNES to take up renewable energy based electrification of 18,000 villages in remote and inaccessible parts of the country by 2012. According to Ministry of Power officials, funds of about Rs 10,000-15,000 crore will be made available to the rural power utilities at 2-2.5 per cent per annum interest rate. With the new Electricity Act and this type of funding it becomes very attractive for micro utilities to come up in rural areas. Thus it is envisaged that a small rural power cooperative can be set up to produce 200-500 KWe of power and supply all the electricity demands of one or two villages. Again this utility can lease the existing SEB power line infrastructure for its purposes.

Accelerated Rural Electrification Programme (AREP)

Government of India in the Budget for 2002-03, has announced the introduction of a new Interest Subsidy Scheme called Accelerated Rural Electrification Programme. With the Interest Subsidy Scheme, States should be able to give this programme the requisite momentum.

An outlay of Rs.164 crores has been provided for this Scheme during 2002-03. The interest subsidy will be at 4% and would be provided to the States for the loans to be taken for rural electrification of un-electrified villages including Dalit Basti.

Kutir Jyoti Scheme

The Government of India in 1988-89 launched a programme called Kutir Jyoti for extending single point light connections to the households of rural families below the poverty line including Harijan and Adivasi families to improve the quality of life of such poor families. Under this programme, one time cost of internal wiring and service connection charges is provided by way of 100% grant to the State Governments/State Electricity Boards through REC The money provided for release of Kutir jyoti connections covers the service line from the pole, the fuse unit, switch, the meter and board. It also covers the cost of single point internal wiring and the cost of bulb.

Keeping in view the current cost of material and labour charges, the Government has now approved to revise the cost from the present Rs.1,000/- to Rs.1800/- per connection in respect of special category States and Rs.1500/ - per connection in other States. Government has also decided that under this Programme, only metered connections should be given. During 2002-03, it was proposed to provide 6, 53,007 connections. The programme, by and large, has been successfully implemented in the States barring a few States like Assam, Goa, J and K, Manipur, Orissa, and Uttar Pradesh and West Bengal. These States have not been able to implement the programme at the desired pace based on performance review upto March ending 2004.

Exploitation of the abundant biomass energy resources available in our country is being accorded a high priority by the MNES. The implementation of projects is being facilitated through comprehensive programmes by the Ministry. The ministry aims to create a favourable policy environment, encourage technology upgradation and ensure market for the power generated. In January 2000, the MITCON (Maharashtra Industrial and Technical Consultancy Organization) was appointed the lead programme partner of the MNES, for the promotion of biomass and bagasse cogeneration on an all-India basis. The mandate under this programme for the next 15 months was to financially close projects equivalent to 200 MW of exportable surplus, including 160 MW from sugar mill cogeneration and 40 MW from other biomass materials. This effort was expected to cover nine major sugar-producing states in the country and also other states with biomass availability.

MNES has also taken up few strategies to improve this sector. 1) Biomass gasifiers capable of producing power from a few KW upto 1 MW capacity have been successfully developed indigenously. 2) Indigenously developed small biomass gasifiers have successfully undergone stringent testing abroad. 3) Biomass Gasifiers are now being exported not only to developing countries of Asia and Latin America, but also to Europe and USA. 4)A large number of installations for providing power to small scale industries and for electrification of a village or group of villages have been undertaken. 5) The Biomass Gasifier Programme has been recasted to bring about better quality and cost effectiveness. 6) The programmes on biomass briquetting and biomass production are being reviewed and a new programme on power production linked to energy plantations on waste lands is proposed to be developed.

The following steps have already undertaken in the **10th five year plan** to facilitate the rural electrification programme (Government of India, 2002).

1. To identifying remote areas where power supply from the conventional grid will be prohibitively expensive and make it a priority to provide off-grid supply from renewables for these areas. Create provisions for integrated generation and distribution of off-grid energy supply.
2. To encourage private sector investments in renewable energy sources by promoting a bidding process for available subsidies. Award contacts to private entrepreneurs who provide maximum benefit with the lowest amount of subsidies.

3. To promote local private sector management of both generation and distribution for off-grid supply from renewable sources.
4. To optimize energy plantation by raising plants on degraded forest and community land.

The above plans, programmes, schemes and initiatives taken at the government level no doubt bear some fruitful results in respect of village electrification. But being a developing country its implementation is not at the desirable space. In the next section, the paper highlights a few successful case studies in respect of rural electrification through biomass gasifier.

3. CASE STUDIES OF BIOMASS GASIFICATION BASED POWER PLANT IN INDIA

A country like India where most of the people is living in the village and normal grid electricity is difficult to access in the remote village then decentralized power generation based on renewable is an attractive option to meet village energy needs. A list of biomass gasifier installations in India is given below:

3.1. Illustrative list of application-wise biomass gasifier installations[*] Power applications

Type of application	Supplied Designed by	Place of installation	Capacity	Biomass feedstock
Captive power	Ankur	Forest Development Corporation, West Bengal	1 x 30 kVA	Woody biomass
		2 Textile units in Gujarat	1 x 120 kW 1x 60 kW	
		4 cold storage units in Uttar Pradesh	3 units with 1 x 100 kW capacity each 1 unit with Ix 200 kW capacity	Rice husk
	IISc	Manufacturer of electrical insulation boards, filter grade paper and allied products, Karnataka	500 kW	solid bio-residue such as mulberry 'stalk
		Navodaya Vidyalaya, Karnataka	100 kW	N.A.
Rural electrification	Ankur	Gosaba island, Sunderbans, West Bengal	5 x 100 kW	Woody biomass
		Chhotomollakhali, Sunderbans, West Bengal	4 x 125 kW	
		Khtrichera, Tripura	4 x 250 kW	
	IISc	Hoshalli village, Karnataka	20kW	Woody biomass
		Ungra village, Karnataka	20kW	N.A.
		Hanumanthanagara, Karnataka	20kW	Woody biomass

3.1. Continued.

Type of application	Supplied Designed by	Place of installation	Capacity	Biomass feedstock
		Port Blair, AandN islands	100 kW	Woody biomass
		Karavatti, Lakshwadeep island (oronosed)	250 kW	N.A.
Other electrification projects	DA/DESI Power/ NET PRO	Desi Power Orchha(P) Ltd, MP	100 kW	Woody biomass
		IIT Delhi, Mechanical Engg	12kW	N.A.
		Desi power,bodhadhara	100kw	
		Dewan estate,karnataka	50kw	
		Desi power,mahanadi,orissa	120KW	
		Devpower corp. tamilnadu	120KW	
		GB Engineering enterprises, Tamilnadu	120KW	
		Desi power,KOSI Ltd. Bihar	50kw	
		Desi power, baharbari Bihar	50kw	
		SKET, Phase I, Kamataka	120 kW	
		SKET, Phase 11, Kamataka	120 kW	
		MVIT, Phase I, Bangalore	120 kW	Woody biomass
		MVIT, Phase 11, Bangalore	120 kW	
		Varlakonda, Kamataka	50kW	N.A.
		GB Food Oils, Tamil Nadu	120 kW	
		V.LT Vellore Tamil Nadu	120 kW	

Thermal applications

Supplied designed by	Place of installation	Capacity	Biomass feedstock
Ankur	CO2 manufacturing, Guiarat	I x 150 kW	Woody biomass
	6 ceramic firms in Gujarat	One firm with I x 300 kW and I x 500 kW units; One firm with I x 300 kW unit; Four firms with 2 x 300 kW units	
	8 firms in Gujarat	One firm with I x 10 kW unit; One firm with I x 20 kW unit; One firm with I x 40 kW unit; One firm with I x 60 kW unit; One firm with I x 100 kW unit; One firm with I x 300 kW unit; One firm with 2 x 300 kW unit; One firm with I x 500 kW unit;	
	Industrial abrasives manufacturing unit, Tamil Nadu	I x 500 kW	
	Firm, Maharashtra	I x 500 kW	
Ankur	Firm, West Bengal	I x 100 kW	Rice husk
IISc	Mis Agro Biochem, Kamataka for marigold flower drying	IMW	Woody biomass

Thermal applications (continued)

Supplied designed by	Place of installation	Capacity	Biomass feedstock
	Thermal Central Building Research Institute, Roorkee, Uttar Pradesh	800 kW	N.A.
	Tea drying, Bangalore	1.5MW	
TERI (through several manufacturers)	CO2 manufacturing, Junagarh, Gujarat	2 x 150 kW	Fire wood
	Manufacture of magnesium chloride, Kharagodha, Gujarat	2 x 150 kW	Firewood
	Silk dyeing, Bangalore	13 x 20 kW	"
	Green brick drying, Palghat, Kerala	I x 20 kW	"
	Rubber drying, Kerala and Tamil Nadu	6xlOOkW	Rubberwood,coconut shells,cashew nut shells
	Silk reeling, Kamataka and Tamilnadu	30xlOkW	Firewood
	Cardamom curing, Sikkim, Bhutan	-150 x 20 kW	"
	Institutional cooking, Beas Satsang, Punjab	I x 100 kW	"
	Cooking in tribal school, Gram Vikas, Berhampur, Orissa	I x 10 kW	"
	Bamboo mat factory, Gram Vikas, Orissa	I x 20 kW	"
	Rice mill, Orissa	I x 20 kW	"
	Drying of mushroom and mahua flowers, Pradan, MP	I x 10 kW	"
	Food processing (Tooty-frooty) factory, Bangalore	I x 20 kW	"
	Melting of Lead in a battery reclamation factory, Bangalore	I x 20 kW	"
	Crematorium, Nagarik Sewa Mandal, Ambemath, Maharashtra.	I x 100 kW	"
	Puffed rice making, Dharwar Kamataka	I x 10 kW	"
	Khoya making, Raiastan	I x 10 kW	,,
	Steel re-rolling, Haryana	1x100kw	,,
AEW	Cooking in hostels, jails (AP. and T.N.)	N.A.	N.A.
	Rubber drying, Kerala	I x 100 kW	
Cosmo	Steel re -rolling, Raipur	Several	
Radhe Industries	Ceramic firms, Gujarat	Several	
Harris	Rubber drying Kerala	6 x 100 kW	

* Sources: (1) Internal documents and in-house publications from institutions such as Ankur, NETPRO, DESI Power, TERI, and lISC, (2) Web-based information: http://ankurscientific.com http://www.desiDower.co,in http://cgpl.iisc.ernet.in, http://www.teriin.org

From the above table it is quite clear that application on rural electrification is very few compared to other applications. Though the potential of biomass in India is enormous still its application is hardly few. It is concentrated in a few states such as Karnataka and West Bengal. However, it is worth mentioning that several biomass projects which are installed and found successful in many respect. Now we shall present some of these successful projects.

1) Hosahalli and Hanumanthanagara Project

The biomass gasifier-based decentralized power generation systems were implemented in Hosahalli and Hanumanthanagara villages of Tumkur district in Karnataka (Somasekhar et al.2000).

Hosahalli was a non-electrified village and Hanumanthanagara was electrified (grid-connected); however, only 30% of the houses were electrified, at the time of project initiation. The number of households in Hosahalli is 35 and Hanumanthanagara 58 (Table 3.2). Hosahalli did not have any pumps or a flourmill. Kerosene-based traditional wick lamps were used for lighting. Women carried water from a polluted open water tank nearly 1 km away from the village. Farmers depended on rainfed agriculture which is subjected to vagaries of monsoon, with low crop yields and occasionally hired diesel engines to pump water for irrigation or partial supply from irrigation tank. The bio-energy project was planned and implemented by CST (The Centre for Sustainable Technologies). Energy forests were established in both the villages; Hosahalli during 1988 and Hanumanthanagara during 1996. The gasifier-based power generation system was installed in Hosahalli during 1988 and in Hanumanthanagara during 1996 and the end-use systems were installed in phases. Local youth were trained to operate and undertake minor maintenance of the systems. CST obtained funds for the implementation of gasifier power system in both the villages. Village committees managed the systems, taking decisions on operation, supervision of the operator, protection of the forest and ensuring payment for the services provided.

Features and Performance: The installed capacity and load for different services are given in Table 3.2. The capacity of biomass gasifier system installed is 20 kW in both the villages. Even though the installed end-use capacity is higher (30 and 37 kW), the load is distributed such that the irrigation load is scheduled for day hours and other load activities are planned for evening hours (6 to 11 p.m.).

In Hosahalli, initially a 3.7 kW power generation system was commissioned in 1988. The 20 kW biomass gasifier system was commissioned in 1997 at Hosahalli and 1996 at Hanumanthanagara. Table 2 provides details regarding the performance of the system and services provided to the village over the last six years. It is important to note that the power generation system was operational in Hosahalli for over 90% of the days during 1998–2003.

The power generation system functioned for over 70% of the days on dual fuel mode, indicating the operation of the gasifier system for power generation using biomass. The system was operated in diesel-only mode for power generation in about 25% of days, due to non availability of processed wood fuel or problems with the gasifier system and sub-elements or non-availability of trained operators. The total electricity generated varied from 12 to 22 MWh per year.

Table 3.2. Features of biomass gasifier systems

Description	Hosahalli	Hanumanthanagara
Year of establishment	1988*	1996
Size of village (number of households)	35	58
Population	218	319
Energy plantation (ha) raised	4	8
Installed capacity(kWe)	20	20
Installed end-use capacity(load)		
Lighting	4.0	4.0
Drinking water	2.6	2.6
Flour mill	5.6	5.6
Irrigation pump	18.5	25.5
Total installed end-use capacity	30.7	37.7

The installed capacity in 1988 was 3.7 kW and was expanded to 20 kW in 1997.
Source: Ravindranath et al., 2004

Table 3.3. Hosahalli: System operation and provision of utility services

System operation and services provided	2003	2002	2001	2000	1999	1998
Days operated during the year	355	358	347	298	343	349
Days on dual fuel mode	287	272	250	162	257	269
Days on diesel mode	68	86	97	136	86	80
Days services provided for lighting during the year	355	349	347	287	310	300
Days services provided for drinking water	353	344	339	293	338	295
Days services provided for flour milling	162	97	155	92	180	125
Days services provided for irrigation water	**79**	**88**	**39**	-	-	-

Source: Ravindranath et al., 2004

The general philosophy adopted in this project implementation was to provide the basic services like piped water and lighting at home and streets in a reliable manner. To achieve this objective, occasionally services are provided using diesel system, if gasifiers cannot be operated. In Hosahalli, lighting and piped drinking water services were provided for over 85% of the days during most years. The flour mill was operated twice or thrice a week depending on the demand for milling of grains. The irrigation system was operated depending on the crops grown, area irrigated, cropping season and demand from farmers. The basic services such as home and street lighting and piped water supply were provided on most days. This is a unique achievement for a village in India. Table 3.4 provides data on electricity generation and biomass and diesel consumption per kWh during 1998–2003 at Hosahalli.

Cost analysis is carried out using only the variable or recurring costs, namely the mean monthly fuel, OandM costs. The fuel, OandM costs at different loads are given in Table 3.5. It is observed that the OandM cost per unit declines from Rs 5.85/kWh at a load of 5 kW to Rs 3.34/kWh at a full load of nearly 20 kW on selected days (2003 cost data).

Table 3.4. Hosahalli: Total annual electricity generation and fuel consumption

Description	2003	2002	2001	2000	1999	1998
Electricity generated kWh/yr in dual fuel mode	18651	17185	12775	7238	9617	9300
Electricity generated kWh/yr in diesel-only mode	3326	3992	3476	5251	3267	2723
Total electricity generated kWh/yr	21977	21557	16251	12489	12884	12023
Average wood consumption rate kg/kWh/ in dual fuel mode	1.8	1.64	2.07	1.28	1.27	1.32
Diesel use in duel fuel mode 1/kwh	0.063	0.077	0.086	0.109	0.173	0.182
Diesel use in diesel-only mode 1/kWh	0.567	0.76	0.779	0.564	0.379	0.432
Diesel substitution in per cent under dual fuel mode	85.55	87.02	80.69	80.67	54.35	58.33

Source: Ravindranath et al., 2004

The biomass-based decentralized power generation system implemented in Hosahalli village has provided multiple social, economic and environmental benefits; some measurable and others not. The potential benefits of large scale spread of decentralized biomass power systems are as follows.

Table 3.5. Fuel consumption, units of electricity generated and O and M costs at different loads Maintenance

Load (kW)	Diesel cost (Rs/h)	Biomass cost (Rs/h)	Engine (Rs/h)	Gasifier (Rs/h)	Labour cost (Rs/h)	Total cost (Rs/h)	Cost/kWh(Rs/kWh)
6.0	16.4	9	5.42	0.98	6.25	38.05	5.85
7.0	21.1	10.5	5.42	0.98	6.25	44.25	4.92
8.5	18.74	10.5	5.42	0.98	6.25	41.81	4.65
11.5	22.26	15	5.42	0.98	6.25	49.91	3.56
15	25.77	18	5.42	0.98	6.25	56.42	3.52
20	42.17	25.5	5.42	0.98	6.25	80.32	3.34

Cost: Wood = Rs 0.75/kg; Diesel+transport = Rs23.45/1; Engine maintenance = Rs 5.42/h; Gasifier maintenance = Rs 0.98/h, Operator wage = Rs 6.25/h.
Source: Ravindranath et al., 2004

Provision of reliable and safe water supply for households is made on most days near the door-step, from a deep borewell. This reduces the drudgery involved in lifting and carrying water from an open water pond nearly a kilometer away. Women in this region spend about 2.6 h/day in collecting water. Further, the quantity of water consumed, an indicator of quality of life, which was low earlier (26 l/capita/day) has gone up based on field observations, due to nearly 2 h of water supply near the door-step of the houses. Women reported improvement in their own health as well as that of their children due to safe water supply and increased use of water and better hygiene. Electricity for lighting in all houses has helped school going children in their studies and women in their household chores. Earlier, women used to walk miles to the neighbouring village for milling grains at least once a week. Now a flourmill has been installed in Hosahalli itself. The unique feature of the project in Hosahalli is equitable sharing of benefits by all the households and reliable provision of services on most days in a year, contributing to improved quality of life for all and counted as social benefit.

The economic benefits include employment and income generation and increased crop production. Establishment of energy forest, harvesting, transportation, wood-fuel chips, preparation and operation of the decentralized power generation system has created employment for two persons on most days and many more during different seasons. In Hosahalli 17 farmers irrigated 20 acres, growing labour-intensive cash crops such as vegetables and mulberry, thus creating employment and generating income.

Environmental impacts are also considered. Raising multi-species energy forest has led to soil and water conservation in the lands subjected to degradation. If mixed species forestry, as done in Hosahalli, is adopted, it will contribute to biodiversity conservation in degraded or wastelands. Biomass is low in ash compared to coal, leading to no or insignificant ash production. Finally, biomass combustion leads to insignificant sulphur emission. In the absence of biomass gasifier-based power supply, Hosahalli village would have used kerosene for lighting and diesel engine for pumping water for irrigation or would have been connected to the centralized grid, where nearly 70% of electricity is generated from coal power plants. Diesel, kerosene or coal combustion leads to emission of CO_2 and other greenhouse gases. In Hosahalli, if all the services were to be provided by diesel-based decentralized generation system, the total diesel consumption for generating 18,900 kWh of electricity annually (average of 2001 and 2002) would be 12,995l. The CO_2 emission from diesel use avoided would be 35 t CO_2/yr. The Hosahalli case study shows the potential of sustainable biomass-based decentralized power generation system for mitigation of greenhouse gases in small villages (< 500 population), which dominate rural India.

Another two successful projects in West Bengal are explained below. West Bengal Renewable Energy Development Agency (WBREDA), Calcutta India has awarded the contract for Electrification of Chhotomollakhali Islands in the Sunderbans using biomass gasifier to our associates in India. This is the second such project with a power plant rating of 500 kW.

2) Chottomollakhali Project

Chottomollakhali Island in Sunderbans is situated in the district of South 24 Parganas is about 130 km away from Kolkata. It is difficult to extend grid electricity to Chhottomollakhali Island due to prohibitive cost involved in crossing of various rivers and creeks. In the absence of electricity, the economic activities of the island were suffering. The Biomass Gasifier based Power Plant (four modules of 125 KW each) on 29th June 2001 has been set up set up by West Bengal Renewable Energy Development Authority (WBREDA) and it is catering to electricity needs of domestic, commercial and industrial users. Four villages (Chhottomollakhali, Taranagar, Kalidaspur and Bodo Mollakhali) of the island are benefited with electricity from the power plant (Mukhopadhyay 2004).

The power plant detail is given in Table 3.6.Average generated power is 400 kW. The generated capacity depends on the local demand factor. It shows that the demand peaks on Monday (450 kW), which is a 'Hat baar' and falls to 350 kW on Sunday.

Table 3.6. Power plant in detail

Lifetime of the plant	15 years
Plant capacity	4x125 kW
Average power generation	400 kW
Internal loss	100%(40kW)
Line loss	5%(20kW)
final selling average	340kW
No. of consumers	225 (industry:1, commercial:74, household:150
Hours of operation	5PM to 11PM
Fuel consumption pattern under full load condition	(a) Biomass (80%) (b) Diesel (20%)
Length of distribution line	HT line-5kms and LT Line-7kms
Substations	4
Name of the manufacturer of gasifier	M/S Ankur Scientific Energy Technologies Ltd.(ASCENT) Baroda, India

Source: Mukhopadhyay, 2004

To run the gasifier plant, woody biomass is being bought from local markets with an average price of Rs. USD 0.02/kg. This BGBPP needs on an average 1-kg woody biomass per unit of power generation (per kWh). The survey period requirement of wood is 400 kgs/day (drywood), which is met by the purchase from the local market. Since the woody biomass content moisture (10–15%) the amount of purchase is higher than 400 kg.

To ensure steady supply of woody biomass for the project and the gasifier programme to be environmentally viable needs to be integrated to the plantation/ afforestation programme to guarantee the source of raw material without a threat of deforestation. It will also help to maintain eco-balance through replantation. WBREDA has initiated the programme and planted quick growing plants like eucalyptus, sirish, tetul and babla (energy plants) in the riverine wasteland.

In future, the wood from these trees will be used as one of the fuels for this plant. This program has also created employment opportunities and villagers are normally employed according to the requirements of the power plant. WBREDA has set up a 10 ha energy plantation in 2000–2002. The average cost to set up this plantation is USD 10235.41. Rotation period is estimated to be 5 years and the average production is 5 t /ha/ year as estimated.

Three types of tariff exist for three categories of consumers, i.e. USD 0.10/kW for industry, USD 0.09/kW for commercial and household are paying USD 0.08/kW in consultation with the village community. Revenue collection is administered by WBREDA. Total average monthly revenue generated varies from USD 511.77–USD 614.12. This amount is utilised in plant operation and maintenan« (OandM) purpose. The deficit expenditure of plant's operation and maintenance is met by Government of West Bengal.

It has been revealed that the gasifier has influenced the economic activities and quality of life of the villagers in a number of positive ways. The benefits generated through the programme over the year are also estimated.

Before the introduction of gasifier power for lighting, the commercial units used to depend on power supply from privately owned a few 10 kW-diesel power generator set. The household sector was completely depended on kerosene and in very rare cases on solar power. Use of kerosene is very inefficient use of fuel in India. It has already a high scarcity value and is very expensive. The additional scarcity value of diesel in the study area arises from the typical geographic location of it by way of high cost and problems in transportation of the fuel in a boat from the nearest mainland canning. The high scarcity value of diesel gets reflected in the high cost of unit of power generated by this privately owned generator supply. The inefficiency of diesel use per unit of power generation was USD 0.49. But the cost of per unit of gasifier-generated power is USD 0.09. Excess payment is an indicator of economic inefficiency. Not only the fuel efficiency of the diesel power generating sets are low they also have many environmental effects in terms of noise and air pollution. The efficiency derived in case of fuel use pattern for cooking and lighting has also been estimated.

The market demand for fuelwood is gradually declining by the domestic users and will benefit a BGBPP in the long run. This reduction in cost of power supply can be assessed from the savings in kerosene, diesel and by pulling out generator set that has been made possible by the gasifier power. In the post-gasifier period inefficient use of diesel generation set has been completely eliminated for the commercial units (Table 3.7).

As none of the commercial units are continuing with the decentralised diesel set using appliance very few of household and commercial sectors are still continuing with their use of kerosene (which is as an input to light the lamp power) as a source to supplement gasifier power to meet the unmet part of power demand from gasifiers. For the domestic sector the supplementary sources are kerosene and solar power. The cost of diesel use per unit of power generation is USD 0.49 but the introduction of the gasifier has reduced it to USD 0.09 per unit. Thus during the post-gasifier period commercial units as well as households are benefited. In terms of simple economic cost it can be said that holding all other components of cost of power generation constant the use of new technology has been successful in bringing down the fuel cost of power from USD 0.49 to USD 0.09 per unit of power generated.

Table 3.7. Fuel uses pattern for lighting

	Commercial		Household	
	Pre-gasifier	Post-gasifier	Pre-gasifier	Post- gasifier
Kerosene	None	None	83	None
Diesel	100	None	None	None
Solar	None	None	17	None
Gasifier	None	100	None	100

Source: Mukhopadhyay 2004

This is an indicator of benefit from the installation of gasifier power. The gain in efficiency has been reflected through gain in consumer's welfare. The beneficiaries have expressed several kinds of benefits, tangible and intangible. The benefits which can be quantified and which have influenced the quality of life is discussed below(Table 3.8 and 3.9).

The quantifiable benefits are those, which emerged due to gain in efficiency in generation, and hence reduction in price per unit of power supply set at the consumer end. The price paid by the consumers in the pre-gasifier period to the diesel generators was much higher and arbitrary. A policy with correct economic incentive will be one, which does not lead to overuse/wasteful use/overstress on existing supply set.

Table 3.8. Benefits of commercial units

Type of benefit		Number of units reporting		% in sample
Savings in monthly exp(USD)				
0.02-1.02		9		18
1.02-2.40		10		20
2.40-3.07		15		30
3.04-4.09		9		18
4.09 and above		7		14
Increase in business hours				
Nil		1		2
<1		6		12
1-2		32		64
>2		11		22
Connected points	Pre-gasifier	Post-gasifier	Pre-gasifier	Post- gasifier
1-4	45	12	90	24
5-10	5	35	10	70
>10		3		6

Source: Mukhopadhyay, 2004

Table 3.9. Benefits of household

Type of benefit	Number of units reporting	% in sample
Savings in monthly exp.(Rs)		
1.02-2.40	16	32
2.04-3.07	18	36
3.07-4.09	11	22
4.09 and above	5	10

Source: Mukhopadhyay, 2004

Essentially an appliance type uniquely represents a particular level of power consumption at certain instants of time. Assuming identical supply quantity at the distributor end a consumer with 100-W bulb consumes 100-W/h of power, while one with a tube consumes 40 W/h of power. But under pre-gasifier regime since the different consumer categories were supposed to pay the uniform price per day both the categories as mentioned above had to pay the same price per day despite different levels of consumption. It is noteworthy that given the lump sum nature of the price charged there was no incentive to conserve power at the consumer end through use of efficient appliances. The pricing structure was unable to distinguish between efficient users and inefficient users giving rise to inefficient system of pricing and appliance choice. Inefficiency of pricing has been removed in the gasifier regime. Per unit price of gasifier power is less than that of diesel power. An economically efficient tariff structure in post-gasifier regime has been successful in encouraging consumers to choose efficient appliances. Although in post-gasifier regime their total units of power consumption has gone up due to more connected points. Still compared to the pre-gasifier their total cost burden is much less. The table above indicates the range of actual monthly savings for beneficiaries. These results from efficient appliance choice by consumers as well as effective reduction in unit cost for gasifier power compared to diesel generated based power.

All commercial units connected to the gasifier supply generate a collective illumination-giving rise to external benefits for the villagers. This has, in fact, given rise to business hours beyond 6:30 PM or even 7:00 PM, which was usually the standard closing time for the pre-gasifier period. Out of the total surveyed commercial units 70% have reported that increased business hour has been productive to them by way of increase in the total sales and hence a more profitable turn over. This qualitative improvement can be quantified in terms of gain in total turn over. Availability of gasifiers power at a cheaper rate has in turn given rise to power consumption through purchase and use of larger number of lighting appliances reflected through their demand for a larger number of connected points compared to pre-gasifier scenario. It shows a significant saving in monthly expenditure of electric bill with an average of USD 2.67 for each commercial unit. At the same time, there has been a considerable increase in business hours and number of connected points leading to an average increase of monthly business turnover of USD 20.47. So on the whole post-gasifier period has brought in a propitious time for the commercial sector of this village island.

Direct qualitative benefits gained by households can be seen from the savings in the monthly expenditure on power. Some of the households could now afford to buy durable consumer goods like radio, T.V, electric iron etc with the savings generated by the installation of BGBPP. They are aware of children's comfort and enhancement of study hours during night. The study concluded that the consumers are saving on an average USD 2.67/month in post-gasifier period. From the field survey it has also been noted that the household consumers are now using on average eight plug points. Prior to gasifier power to save kerosene or diesel power, nighttime activities of the villagers were almost non-existent. Now with gasifier power available upto 11:00 PM., usël household activities like watching T.V, socialisation, etc. have increased and thus improved the quality of life. The households are aware of the benefits and are demanding longer hours of service from the BGBPP.

In this connection we have to mention that the benefit (or monthly saving) incurred from the BGBPP by the household will help them to switch over from traditional chullah to kerosene/LPG type. More specifically, beneficiaries are now willing to spend the extra amount they have saved from electricity on the efficient type of fuel for cooking. Overall this benefit will indirectly save time and health costs for households.

A huge potential demand for BGBPP power has been observed. Moreover problems also have been identified due to limited hours of power supply. All the surveyed beneficiaries and the non-beneficiaries have expressed their willingness to pay the existing rate of tariff despite their awareness that it is 2.5 times higher than conventional grid power tariff of West Bengal. Fifty percent of the surveyed non-beneficiaries have shown interest to take up the BGBPP lines but the lines have not yet been extended due to transmission and distribution constraints.

A huge demand from potential consumers has also been experienced. The further extension of transmission and distribution would be justifiable given the density of settlements in the neighbouring areas. It was assessed that with increasing expansion of the TandD network unmet demand would be minimized in the process. It is found that all households and commercial beneficiaries would like to receive power increased from 6 to 24

h. This increased power supply will lead to more profitable turn over for the commercial sector. Though the average income of the household is not high still they are willing to pay more because it is expected that increased power supply will help the students to study for more hours comfortably and use of more electric appliances leading to further improvement of quality of life for the villagers.

The findings of the study (Mukhopadhyay, 2004) indicate that the BGBPP has made a very positive impact on the life of the villagers of Chottomollakhali Island. This has led to increased economic activities and more profitable turn over for the commercial consumers and improved quality of life for the household sector. The consumers now are gaining in terms of lesser electric bill expenditure. All of them have shown a willingness to pay a higher price to get 24 h of power supply. From the BCA(benefit cost analysis) it is evident that the plant can recover its full investment after 7 years. It has also been observed that the BCR is greater than 1 (1.68). Moreover the IRR is 19%, which is more than the cost of capital (8%). Assessing above criteria we may conclude that the project is economically viable. Though the environmental benefits have not been measured in the BCA but the study finds that the BGBPP has a positive environmental impact. But the environmental awareness is very poor among the villagers.

3) Gosaba Project

The first rural electrification demonstration project in the country considered to be a success is at Gosaba, an island of about 156 sq. kms. area in the Sunderbans area in the state of West Bengal in 1997 (Akshay Urja, 2005). Gosaba was selected as a site for rural electrification based on decentralized supply sources, as this was the only option for this region. In addition, the area also has an abundance of biomass resources. The project was implemented by WBREDA in association with MNES, Sunderban Development Department (the local development body), Forest Department and South 24 Parganas Zilla Parishad (local administration). MNES subsidized 75 percent of the project cost and state government gave the remaining. The state electricity board has set up the distribution network, with financing from WBREDA.

The total electricity generation capacity is 500 kW, with five individual gasifier-based units of 100 kW capacity each. The gasifiers are closed-top downdraught systems based on woody

biomass, supplied by Ankur. The plant has two dual-fuel engines that are synchronized with the system and can be operated in parallel. The entire project cost was Rs. 10 million, including setting up of TandD network. The investment for the distribution network in Gosaba amounted to Rs. 1.8 million. The capital cost for the gasifier installation approximates Rs. 25 million per MW. The transmission and distribution line spans over a length of 6.25 kms of high-tension lines and 13.67 kms of low-tension lines, with a cost of around Rs. 175,000 per km. The electricity generation in the plant is at 400V. Within the TandD network, around 45 to 50 consumers are connected every kilometer. The average TandD losses are only 4 percent.

At present around 900 consumers are being provided with power 16 hours a day. In the initial stages of the project, a single 100 kW gasifier unit was installed as the existing load at that time was just around 10 to 20 kW. The villagers were reluctant to participate and there were only around 25 consumers of power. It took some time to convince a local people of the potential benefits and the load growth took around a year. The operating load of the system is 300 kW. Therefore, a maximum of three gasifier units are operating at one point of time and the other two units are kept as standby.

The average daily generation is 950 units over the period of 16 operating hours. The tariff for domestic consumers is at Rs. 5/kWh, for commercial shops and establishments it is Rs. 5.50/kWh and for industrial consumers Rs. 6/kWh. The average household consumption is in the range of 1 to 3 units per day. The households with electricity supply connection have to pay a fixed charge of Rs. 75 per month in addition to the variable charges for the units consumed. The monthly revenue generation at Gosaba is around Rs. 160,000. There have been no defaulters in payment of electricity bills and no electricity thefts are reported. The charges are affordable to the users and they are willing to pay the price for reliable electricity supply. Before the Gosaba project, people in this area were paying around Rs. 9/kWh for diesel-based generation. The state nodal agency's assessment is that demand for electricity among villagers is increasing steadily as more local industries are coming up.

The plant is run by a local co-operative society, which receives funds from WBREDA. This cooperative, which is responsible for ensuring biomass supply, daily plant operation and maintenance, and financial record keeping. For undertaking renovation, repair and maintenance of plants, around 75 percent of the financing comes from the co-operative and the rest from MNES. The success in running this rural energy co-operative is partly attributed to the history of success in co-operative movements in that area. Ankur undertook turnkey operation for the project and at present intervenes in major maintenance and retrofitting functions. It has trained local people in plant operation and maintenance. It also periodically reviews plant operation. One of its service engineers based in eastern India supervises these activities. The state nodal agency, WBREDA, functions as the Technical Backup Unit for the project. It provides both technical and non-technical support for running the system, monitors the operation of the plant and performance of the plant personnel. It periodically conducts tests for the plant operators to monitor their performance.

The socio-economic development of Gosaba after electrification has been immense. The availability of electricity has allowed students to study at night. Small-scale industries (for example lathe machine units, boat-repairing works, and grill welding , domestic iron implements sharpening machines and machines to grind spices like chilli and turmeric, using automated electricity-operated machinery) have been established in the region. Establishment of Xerox machines in shops has facilitated photocopy of costly study books and to make copies of important documents like deeds. An opera÷Áon theatre has been made functional in the Government Health Centre in the Island. With the availability of refrigerators it has become possible to store life-saving vaccines or medicines. Water from ponds is being drawn with the help of electricity- operated pump motors for the purpose of irrigation and cultivation of crops Availability of electricity has given way to entertainment. People are able to watch sports and other programmes on cable television, which were not thought to be possible earlier. Cinema shows are being organized by the local video parlours. A computer-training institute has started in Gosaba (Akshay Urja, 2005). Nowadays people need not go to Canning for any printing press related jobs, or sizing the lens of spectacles. Time is being saved, and the expenditure and risk of travelling has reduced. Electric sewing machines are being used to manufacture fishing nets for the fishermen. Tailoring shops are using electric press. The local shopkeepers and traders have been able to decorate their shops and establishments through strategic illuminations to attract customers. Presently, the income levels have also increased due to increase of sales. Electronic goods shops have proliferated in the region where all kinds of electrical goods are available. Many telephone booths have been established. The scope of work has also increased for people of the region, and there has been a huge improvement in the communications system. The use of wood-generated electricity has upgraded manpower resources, and led to an overall development of the region. The biomass gasifier power plant has been a boon for the people of Gosaba. It has brought about an overall development in their lives.

The villagers have really benefited from the above three power plants located in southern and eastern part of India. In this context the paper also evaluates the two power plants located in northern part of India.

4) Producer-Gas Electricity for Six Villages in North India

A village electrification programme using biomass gasification has been implemented in a cluster of six villages near Kotdwara (Garhwal region, North India), by the Himalayan Environmental Studies and Conservation Organization and the Indian Institute of Technology (IIT), Delhi. A brief review of the biomass energy project in a cluster of villages around Kumbichar village based on (Ravindranath et al., 1995). A 3.5 kW gasifier was installed in Kumbichar village, which uses mainly *Lantana camara*, a wild shrub growing in the village, as feedstock. It is a small village and all 18 houses are provided with electricity for lighting form a 3.5 kW system. During the day, the system is used to operate a flour mill, spice grinding, and a leaf plate making machine. The system is operated by trained village youth. Each household is to pay Rs.20/month for lighting. The operational cost of producer-gas electricity generated was Rs0.93/kWh compared to the grid electricity rate of Rs1.38/kWh (1992 price). A diesel substitution of 70 per cent has been achieved. The system has been expanded to five more villages.

An interesting aspect of the project is the use of the relatively low-density shrub Lantana camara. Lantana grows wild in many parts of India and thus using it for power generation in this demonstration project provides an excellent example of using such a hardy, quick-growing, and copping weed for economic gain. It is also relevant that the households collect and cut Lantana sticks for gasification each day, thus avoiding the growing and collection of wood. This contribution of the village community demonstrates their commitment to the programme and reduces the investment and operational costs.

5) Nari Project

Work done at the Nimbkar Agricultural Research Institute in Phaltan, Maharashtra, has shown that each taluka in the Phaltan district produces enough agricultural residues so that all its electricity demands can be met by using them in 10-20 mw biomass-based power plants (Rajvanshi,1995). The NARI study also showed that besides providing power, the taluka energy self-sufficiency plan could also create 30,000 jobs/year. With the new Electricity Act, taluka energy self-sufficiency can become a reality since the utility can produce and supply power to its customers without the need to go through SEBs. The taluka utility company can also lease the existing transmission and distribution infrastructure of SEBs so that it need not invest in developing its own. This will also help the SEBs to get regular income from their infrastructure. The NARI study also showed that the taluka energy programme could produce Rs 100 crore/year wealth for its inhabitants in terms of biomass production and setting up of new electricity-based industries. With about 3,500 talukas in the country it is therefore possible to produce about Rs 3, 50,000 crore/year extra wealth through the taluka programme. NARI has recently suggested this concept to the Maharashtra Electricity Regulatory Commission. It is also envisaged that electric cooperatives may function on the lines of TV cable operators in rural areas. However, for small power packs of 500 KWe and less to function smoothly in rural areas it is necessary that they be powered by fuel from locally available resources. Thus there is a need to do sophisticated RandD in producing biofuels from renewable energy sources. These biofuels can easily power the existing diesel gensets. Development of liquid fuels like ethanol and bio-diesel from multipurpose crops should be done so that the issue of food and fuel from the same piece of land is taken care of. This will help in creating fuel supply network for small rural based utilities. Besides it will create wealth in the rural areas by producing value-added items like liquid fuel from agricultural residues.

Apart from the rural electrification there are several other successful stories of biomass gasification which can also be mentioned in this respect.

a. The Desi Power Unit at Orchha

The first DESI Power plant was installed in 1996 at Orchha, near Jhansi, a five hour train ride south of Delhi. This plant has now been operating for more than three years (SEI, 1999). It is located at and supplies power to TARAgram, a campus where TARA carries out research, demonstration, training and production activities using appropriate technology. Although TARAgram has acquired a grid connection, the bulk of its electricity requirements are still met by the DESI Power plant. As for all DESI Power stations, a careful site analysis and feasibility report was prepared for the Orchha plant, including an assessment of electricity demand, availability of renewable fuels and existence of interested partners for setting up an IRPP.

On the basis of the study regarding current energy sources purchasing power of households small business opportunities in the region, accessibility of local investment capital and availability of skilled labour and availability of local biomass and 80 kW power station was ordered in December 1995. The DESI Power Orchha unit, located at TARAgram, went into operation in April 1996. The capital cost of the station was Rs. 22 Lakhs. Initially, the sole client was TARA. Ipomea was used as the fuel. An 80 kW plant needs almost one tonne of biomass fuel every day. Managing biomass resources properly is crucial for reliable plant operation.

More than ten local families have created new livelihoods through harvesting, chopping and transporting the weed and delivering it to the power station, where it is sun-dried before being fed into the gasifier. However, it is estimated that the cost of fuel is adding some 20% more than it should to the final cost of electricity produced. The current indications are that the quantity of biomass accessible to the plant in its first year should be at least 50% greater than the long-run annual requirements.

The plant is satisfactorily operated and maintained by local personnel. Local people, with fairly minimal education is being trained and employed for the tasks. Further employment of over 100 persons was created through the factories that purchase electricity from the Orchha plant. Over the past three years, the gasifier has logged about 5,000 hours of operation, running nearly 10-12 hours per day during normal operation. The percentage of diesel replaced by gas depends on the plant load factor. It has been recorded as high as 85%, with the average at about 75 per cent. Competitive pricing of electricity will require DESI Power stations to maintain an average of well above 80%. Plant load factor is a critical parameter to which the economics of power generation is extremely sensitive. The plant load factor is important not only for spreading out the fixed costs, but for raising the diesel replacement rate, reducing the use of costly diesel fuel. The breakeven plant load factor for the Orchha plant is between 50% and 60%. Even with 40% load factor and high biomass cost, the cost of electricity has been Rs.4.0 to Rs 4.5 per kWH, which is still competitive with electricity from the grid.

The Orchha plant has shown that the cost of electricity can be highly competitive with conventional systems, provided that investment capital is obtained on terms similar to those available to larger Independent Power Producers. The two most sensitive determinants of production cost are the cost of biomass fuel and the plant load factor. Efforts are being made to raise the Plant Load Factor at Orchha by adding new clients and industries with energy-intensive applications, reducing the high fluctuations in some of the loads staggering certain activities.

b. Valli Chlorate Company at Kovilpatti, Tamil Nadu

Valli Chlorate Company has been chosen as a successful case study by installing 100% producer gas based gasifier. The company has also increased its profitability by installing the gasifier by reducing cost of energy purchased from TNEB(ITCOT,2005). The company sells Potassium Chlorate at a rate of Rs. 36 per kg. The production cost with EB Power Supply was around Rs. 34 to 35 per kg results in a meager profit margin. Due to gasifier installation, the production cost of Potassium chlorate has come down to Rs. 28 to Rs. 30. This has helped the plant in improving their profits. The operation of the unit is for 8 hours a day and extends some times to 10 hours a day and six days a week.

c. M. Vishveswaraiah Institute of Technology Installation at Bangalore, Karnataka

M. Vishveswaraiah Institute of Technology (MVIT) has sought the assistance of Netpro Renewable Energy Private Limited, Bangalore to install a Dual Fuel gasifier-based power plant in their college premises and to supply power to the college. In the educational institutions category, MVIT is the oldest gasifier unit operating successfully for the last three years. The unit also operates for longer period in a year and has generated more units when compared to other units. Netpro is presently successful in providing captive power to the college to the greatest satisfaction.

d. Bagavathi Bio Energy Limited Installation at Coimbatore, Tamil Nadu

In the industry sector, the unit that is successful in implementing the gasifier program is Bagavathi Bio Energy installation at Mettupalayam. The company has successfully been running as a separate entity to serve the energy demand of a Textile bleaching unit by installing the system in their premises without any major problem (ITCOT,2005).

Though other units covered by the study were also operating the gasifiers, the data available from the units other than the above four is scanty and further, units were also not willing to share the real performance of the gasifiers. Some units found the operation not economical on dual mode due to high cost of diesel, and hence, gasifiers are not run.

e. PSG College of Technology, Coimbatore

through Ministry of Non-Conventional Energy Sources (MNES) New Delhi implemented a 1x 100kW biomass gasifier on electrical mode at the PSG foundry campus, Neelambur to partially meet the foundry electrical energy requirement and also to promote gasifier system based academic research projects at the college. The plant is commissioned in 30th July 2004(ITCOT, 2005).

f. Jagat Alloys Private Limited (JAPL), Tamil Nadu (JAPL)

has set up only 1 x 500 kW gasifier to meet the load of Ferro Alloy Plant. The plant is commissioned on July 4, 2001 at Nellithurai, Coimbatore district. The project is commissioned to meet the energy requirement of the raw water pumping station at Nellithurai Municipality and the lighting of the surrounding area. The total power requirement is 3 HP and 2 HP for water pumps and a lighting load of 3 kW(ITCOT,2005).

g. Vellore Institute of Technology (VIT), Tamil Nadu

has installed a 90 kWe Gasifier for generating power for captive consumption in the hostels and to carryout R and D activities in the field of Renewable Energy. VIT conveyed that overall performance of the gasifier was satisfactory (ITCOT, 2005).

h. Periyar Maniammai College of Technology for Women (PMCTW), Tamil Nadu

has installed 100 kW Dual Fuel Mode Gasifier system during 2001 and later installed 200 kW 100% producer gas based Gasifier system during 2004. The college has installed the gasifiers for captive consumption and for carrying out RandD activities in the field of renewable energy. The dual fuel based gasifier plant was installed during 2001 and the 100% producer gas based plant was installed on 24th June 2004(ITCOT,2005).

From the above case studies it is reflected how far the village people get benefited from the rural electrification programme. It also gives us a good picture about the pros and cons of biomass gasifier power plant. The attempts so far initiated to install BGBPP in India are mostly with other purposes without implementing rural electrification programme intensively. Moreover, according to the statistics most unelectrified villages are in Bihar, Rajasthan and Madhya Pradesh.

4. ASSESSMENT OF THE CASE STUDIES

The case studies discussed above predict the feasibility of the biomass gasifier based power plant in rural remote villages in India. Several benefits have been derived from the proper implementation of the gasifier plant. The Hosahalli and Hanumanthanagara, case study has demonstrated the technical and operational feasibility of a decentralized biomass gasifier-based power generation system to meet all the rural electricity needs in an environmentally sustainable manner. It is one of the oldest applications of rural electrification programme. Another feasible option is Chottomollakhali situated in eastern part of India indicates that

BGBPP has made a very positive impact on the life of the villagers of Island. This has led to increased economic activities and more profitable turnover for the commercial consumers and improves quality of life for the household sector. Further, Gosaba Island installation has also been a successful case study for the reason that the energy generated from gasifier has turned the socio economic status of the Island along with community development. Even when the cost of energy generation is high, the energy generated from gasifier has played a pivotal role in improving the life style of the rural masses of Sunderbans, where grid penetration is impossible.

Several other feasibility studies(BERI, ITCOT,2005) on biomass gasifier based power plant not exactly for rural electrification but dealt with the objectives like water pumps system, cooling machine, rubber drying, cardamom drying, silk reeling machine for convenience of the village activities. The above application of the biomass power plant also helps to generate employment opportunities and reduce poverty in the village. But some of these villages are not yet electrified, though there are biomass gasifier activities running for other industrial purposes. In our opinion these villages should be first in the targeted list for electrification. It will be easier also to introduce the biomass plant in that area.

Whatever the programmes so far taken by MNES and other state governments regarding electrification through gasifier , mostly concentrates on southern part of India and }¶w in eastern part. So the regional variation is a dominant factor in this regard.

Moreover the attempts of biomass gasifier plant in southern part of India mostly applied on small and medium scale industries. So there are rooms for trading off between rural electrification and Small and Medium Enterprises facility. These vital issues need to be addressed by the Government of India according to priority basis.

Further, forestry in degraded lands is a well-known climate mitigation option to sequester carbon in soil and standing trees. Biomass power based on sustainable biomass supply is a climate mitigation option for substituting fossil fuels (coal used in power stations, kerosene and diesel), leading to reduction in the emission of greenhouse gases. This is a very positive contribution from the environmental point of view.

However the most important issue is cost related. There is a general perception that the cost of electricity generated by renewable energy technologies is always higher than electricity generated by fossil fuel sources. While this is true in many places, it is no longer valid for the rural areas of l India. In most of the Indian villages diesel generators are often the only source of power but power from biomass gasifier based plants are considerably cheaper where ever biomass is available. Even for dual fuel operation where 20 % diesel is used, the generation costs are lower, especially with high running hours and loads. The savings are high when pure gas engines are used. Even when grid power is available, the actual cost of power at the point of consumption is very high largely due to line losses in transmission and distribution. High subsidies and financial losses keep the power price low for agricultural pumps but now that industrial and commercial consumers pay the actual cost of power, the biomass gasification based electricity can easily compete when pure gas engines are used.

Table 4.1 explains the capital cost differences among the renewables. It clearly shows the cost advantages in most cases. Financial Advantages of Decentralised Biomass Power Plants here are assessed by the DESI power in a separate manner. As the experience of DESI Power's EmPower partnership programme shows, grid supply to remote areas is not competitive with electricity supply from modern decentralised renewable energy power plants (table 4.2).

Table 4.1. Capital Costs and the Typical Cost of Generated Electricity from the Renewable Options

Sl No. Source	Capital cost (Crores of Rs/MW)	Estimated cost of Generation per Unit (Rs./kWh)	Total installed Capacity(MW)
1. Small Hydro-Power	5.00-6.00	1.50-2.50	1601.62
2. Wind Power	4.00-5.00	2.00-3.00	2483.00
3. Bio-mass Power	4.00	2.50-3.50	234.43
4. Bagasse Cogeneration	3.5	2.50-3.00	379.00
5. Bio-mass Gasifier	1.94	2.50-3.50	60.20
6. Solar Photovoltaic	26.5	15.00-20.00	2.54
7. Energy from Waste	2.50-10.0	2.50-7.50	41.43

Source: Planning Commission, 2005

Table 4.2. COST OF SUPPLYING POWER TO A VILLAGE

Generation MW TandD Losses End Use Energy						
	MW	Cost Rs	MW	Cost Rs	MW	Cost Rs/MW
Centralised Grid Supply	1	35 million	0.3	5million	0.7	57 million
Decentralised biomass power plant (gasification)	1	35 million	0.1	5 million	0.9	44 million
SAVINGS: Decentralised Vs. Centralised						

	Power	Power Saving	Avoided Cost	Saving in Cost	Amount
Generation/End Use	0.2MW	22%	13million Rs/MW	29.5%	
CO_2 emissions					5500t/y per MW

Source: Desi Power (2004)

Though the above explanation gives us a fair idea about the cost differences between decentralized and centralized power system and also several benefits which can be derived from the Biomass plant still the negative side cannot be ignored. The advantages and problems exclusively from the BGBPP are outlined below.

1) In India biomass technology is mature with several designs and manufacturers who undertake planning and commissioning of small-scale biomass power systems and who also provide performance guarantee. 2) Biomass gasifiers are available in different capacities for decentralized applications from 5, 20, 100 to 500 kW in India. 3) Biomass gasifier-based system can be operated from 1 to 24 h a day, depending on the load. The system can be operated 365 days in a year, if needed. Woody biomass feedstock can be transported over shorter distances and stored. 4) Biomass gasifiers can be installed and operated in any village where biomass is available or can be grown, except probably in desert areas. Such flexibility

does not exist for other renewables such as solar, wind, micro-hydro and biogas systems.5) Biomass gasifier technology is indigenously developed and transferred to manufacturers. Maintenance, spare-part supply and servicing facility and infrastructure are available or can be organized. 6) The economic viability is yet to be proven for renewables in India based on monitoring of field-based systems. Preliminary assessments available show that biomass gasifiers are economically feasible and have lower cost per kilowatt hour compared to other energy technologies. 7) Biomass gasifier-based power generation systems create jobs and skills in rural areas in biomass feedstock production, transportation and processing, and in operation and maintenance of the gasifier–engine–genset systems as well as end-use systems. 8) India has vast degraded or wastelands (over 60 million ha), which urgently require revegetation to prevent further degradation. Biomass production, as feedstock for power generation, provides economic incentive to re-vegetate wastelands with energy forests.9) Appropriate guidelines to discourage monoculture plantations and incentives to promote mixed species forestry, with appropriate density will promote biodiversity in degraded lands. Use of forestlands to supply biomass or conversion of natural forest to energy plantations is not desirable and banned in India. 10) In India, coal-based power plants account for 70% of electricity generated. Thus, every kilowatt hour of electricity generated based on sustainable biomass supply, leads to reduction in CO_2 emission by 1.0 kg. Biomass power is recognized as a prime option with high potential for reduction in carbon emission and climate change mitigation.

Despite the above advantages the rate of spread of biomass- based power generation technology in India is low, due to a number of policies, institutional and financial barriers. There is need to address such barriers to promote biomass power in India.

Problems Encountered from the BGBPP

A decentralized power generation system, implemented in a village could face a number of technical, social, political and financial problems over the period of its life. Here, some of the broad problems encountered particularly focusing on the performance of the plant are presented.

Technical Problems

With respect to the technology package, the two critical components are the gasifier and diesel engine. These include the failure of material in the top shell (twice), filter replacement and grate repairs. The gasifier maintenance-related problems which related to the operational problems, use of moist fuel, wrong size fuel, etc. affecting the quality of gas. Engine-related problems which are related partly to gasifier operation and also to the failure of the radiator. Electrical system of the alternator failed twice. These give an indication about the reliability of different components of the system. Some of the gasifier-related problems have been addressed by the RandD group. The problem related to the reactor material has now been resolved using high temperature ceramics and also the filter failure using fabric filters.

Input Supply

The utility does not have adequate storage facility. Thus during rainy seasons the cutting and drying of wood is a problem. The labour availability for harvesting wood in the energy forest and then chopping to the size needed is limited, particularly during the peak crop season. The power generation system was operated on diesel mode during such days, when cut and dried wood chips were not available. The nearest diesel source is about 30 km away and often diesel is not available or accessible to the operator. The system could be operated on days when diesel was not available. The mitigation measures for sustained biomass feedstock supply include: creation of storage capacity, mechanical system for cutting wood to the desired size and drying of wood using the exhaust gas of the engine. The ultimate solution to the diesel problem is the adoption of a 'gas engine', which is at an advanced field testing stage.

Non Availability of Operator

Due to financial constaints fully trained operators necessary for the running of the plants could not be employed in most cases.

Social Problems

The social problems include some disagreement among the members of the management committee or political rivalry. Social problems-related disruption was infrequent; this is a tribute to the most of the village community. The other social problems like unauthorized removal of trees from the forest occasionally, grazing of livestock in the forest, attempted encroachment of the forest land and opposition to the operator from one section of the village community. It is no surprise that the village communities in India are divided over political affiliations, castes, land ownership, etc. But what is surprising is that despite the social and political problems and lack of their understanding by the most of the project team, the decentralized system has functioned uninterruptedly.

Subsidy and Government Policy Related Problems

While the government program has relied mainly on subsidies and an orientation towards target fulfillment, it has had little emphasis on performance. Furthermore, the implementation of the government effort has not been driven by need assessment and performance evaluation. In fact, there has been little emphasis on systematic review of the program. There are also distortions in subsidy policies in terms of the structure and nature of subsidies. Some of these arise due to subsidies being applicable even for commercially viable applications of gasifiers such as productive uses in industries and region-wise (higher subsidy to locations in the north-eastern regions) and category-wise (higher level of subsidy offered to consumer categories belonging to certain socio-economic classes) classification of subsidies. Frequently changing government policy guidelines with respect to subsidies also results in awareness problems among users. The installations of systems are often driven by subsidy motives and draw little commitment from users. There has been no shift from capital subsidy to performance-based incentives such as soft loans and tax credits.

Bureaucracy

The procedures for government subsidy approval and disbursement are lengthy and cumbersome that deters potential beneficiaries (although to be fair, this is not a particular problem for the renewables areas only). A bottom-up structure exists related to project development and implementation that leads to high cost and time overruns due to factors such as procedural bottlenecks and approval needs from multiple agencies.

Finally, there are also adverse impacts due to uneven support to RandD institutions and sudden withdrawals in government support without alternate support in place. For example, after the dismantling of government support for GARPs, there remains no agency for testing and certification of gasifiers. There are uncertainties with respect to resumption of these activities that adversely affects dissemination.

Thus,, an overall assessment lead us to state that the Gasifier based Power Generating systems are technically feasible and financially more viable, when run on 100% producer gas mode and not on dual fuel mode. The project would be more successful if the availability and cost of biomass used in the system are kept at constant check. Energy Plantation would be the most feasible and reliable option for adequate and cheaper biomass availability. Constant Load operation and increased Plant Load Factor (PLF) would further improve the financial viability of the Project.

5. Conclusion and Policy Options

The objective of the present paper is to evaluate the rural electrification programme in India undertaken by the Ministry of Non Conventional Energy Sources (MNES), Government of India, through biomass gasifier power plant. Among the renewable energy technologies, biomass gasifier-based decentralized power generation holds great promise for meeting rural energy needs. A number of case studies have been presented and these case studies more or less establish the feasibility of the gasifier from various fronts. On the one hand it provides commercial, social and environmental benefit, on the other hand it helps to achieve better quality of life through creation of jobs in the power stations, small-scale business, commerce and industries.

We have discussed that the Government of India had already undertaken a number of strategies and policies for rural electrification programme. Due to more feasible option, biomass as considered, Government has taken one step further initiatives in this regard. Among the various programme, the Rajiv Gandhi Gram Vidyutikaran Yojana (RGGVY), Pradhan Mantra Gramodaya Yojana, Distributive Generation, Kutir Jyoti Programme deserve mention.

The biggest challenge in this regard is to electrify 62,000 villages through grid power, during the 10th Five-Year Plan (2002-07) under the Pradhan Mantri Gramodhaya Yojna. The Centre hopes to achieve 100 per cent village electrification in the current Plan. Another 18,000 remote villages will be electrified through decentralised plants based on biomass, gasification of biomass, hydel power, solar thermal power etc. We have also discussed the problems related to biomass gasifier rather than rural electrification. Unless we introduce proper policies and implement them, our objective will not be fulfilled.

Towards this end the paper is suggesting a few policy options.

Suggested Policies

The paper suggests three steps policy suggestions. First the broad policies; secondly a few subsidy oriented policies and finally some specific policies have been suggested.

Broad Policies

Renewable energy including biomass may need special policies to encourage them. This should be done for a well-defined period or up to a well-defined limit and should be done in a way that encourages outcomes and not just outlays. The programme can be accelerated if the support of the government and the budgeted public funds were used to leverage local, private and corporate sector investments in these rural projects.

A policy framework can be established for utilising sanctioned funds earmarked for renewable energy based rural electrification as well as for other rural development programmes (e.g., for kerosene, small scale industries, job creation schemes, etc,) in a more focused and integrated manner.

An annual renewable energy report should be published providing details of actual performance of different renewable technologies at the state and national level. This would include actual energy supplied from different renewable options, availability, actual costs, operating and maintenance problems. The monitoring should also encompass other parameters like user profile (in order to ensure that government support is indeed going to poor households), livelihood outcomes such as increased income, improved food security and gender impacts. In addition to monitoring the performance of devices the assessment should critically review the programme objective and the strategy adopted and suggest course corrections as required. Information on any system that is receiving government support should be made publicly available. It is essential to ensure that independent assessment of performance is done for all renewable projects receiving Government funding. This help in tracking programmes, repeating mistakes and providing mid-course correction.

Apart from the facilities of existing Technology Business Incubators being set up by the DST energy entrepreneurs for renewable energy, energy efficiency, or rural energy also need finance. Financial institutions should be encouraged to set-up venture capital funds for energy entrepreneurs.

Subsidy Oriented Policies

It is well known that **capital subsidy** plays a very important role for rural electrification in India. A critical issue in distributed generation for rural electrification is the cost recovery and the implementation mechanism. Different policy experiments for implementation of DG appropriate technology option (biogas, bio-mass gasification) for their village. For isolated systems it is beneficial to link the DG system to an industrial load (cold storage, oil mill etc.) to improve its load factor and hence it's economic viability. The **capital subsidy** should be based on the annual generation. This is preferably in the form of an annualised subsidy to be provided on actual generation. These pilot projects can be set by panchayats, independent power producer or renewable energy service companies. A mechanism of bidding can be used to obtain the annualised subsidy level. For example, if it is decided to electrify a village using a dedicated producer gas engine and bio-mass gasifier, bids may be obtained from suppliers for the support required annually for the first three years per kWh of actual generation. The project would be given to the lowest bidder. This would require actual tracking of annual generation. This is feasible using existing technologies of remote monitoring and would add only incrementally to the system cost.

A premium on feed-in tariff may not benefit a stand alone plant in a remote area. For such a plant, capital subsidy may be required. Such **capital subsidy** can be linked however, to the amount of power actually generated, if it is given in the form of Tradable Tax Rebate Certificates (TTRC). The rebate claim becomes payable when electricity is generated and linked to the amount of electricity generated. This will also encourage exploitation of better wind sites earlier. The need to keep the certificates tradable arises from the possibility that small generators may not have adequate taxable income to benefit fully from tax rebate. In areas where there is no electricity grid, there should be minimum clearances/permissions required for setting up a Distributed Generation (DG) system. Supply companies/entrepreneurs should be free to set-up micro-grids and recover revenues from customers. This is already provided for in the electricity act. Each state should clearly define guidelines to facilitate this.

Price subsidy for renewables especially for biomass may be justified for several grounds. A renewable energy source may be environmentally benign. It may be locally available making it possible to supply energy earlier than when a centralised system can do. It may provide employment and livelihood to the poor. The environmental subsidy for renewables should be financed by a cess on renewables and fuels causing environmental damage. Price subsidy should be linked to outcomes. For grid connected renewables like biomass Regulatory Commissions (RC's) should provide feed laws to permit renewables to supply electricity to the grid. RC's should ensure that renewables are given a tariff equal to the avoided cost of generation.

Specific Policies

The following specific **policies are to promote rural electrification through BGBPP**.

Fuelwood Plantation: Cooperatives should be encouraged and facilitated to grow tree plantations in villages. Cooperatives which are open to all members of the community and which are non-discriminatory should be given government land on long-term lease. Women should be encouraged to set-up and manage such plantations so that the time they now spend in gathering fuel can be spent productively in a way that empowers them. They should also be provided finance. If organised and managed properly, such plantations are economic and successful as shown by the experience of National Tree Growers Cooperatives Federation, Parikh et al (1997). Field based NGOs could also be involved in this activity. To encourage large-scale plantations, contract farming should be facilitated.

Electricity from Wood Gasification: This can provide electricity based on gasification of wood and can be very useful especially in remote villages. Some institutional arrangements to promote renewable energy are needed:

Finally, for the rural electricity supply mission to succeed it is necessary that close cooperation between corporate sector, government and NGOs is needed. The corporate sector can provide the necessary technological and managerial support, NGOs can create the necessary trust in such utilities and Government of India can help provide soft financing through its many rural development programmes. An energy self-sufficient and hence prosperous rural India will be the first step in making us a developed nation.

Thus the paper gives us a glimpse of hope for the future development of the rural sector in India through sustainable biomass power. There is need for large-scale demonstration in different parts of India along with policy, institutional and financial support for large-scale spread of environmentally sustainable biomass power systems in India and other developing countries. The State as well as central govt. should be more pro active in that respect to install Biomass gaisifier power plant in those un-electrified villages with the help of MNES.

These successful efforts in India briefed in section 3 are really lessons for other countries of this tropical region like Cambodia, Laos, Burma, Vietnam, Thailand, China, Malaysia, Philippine and Indonesia. As we know that the interest in biomass energy is growing in most Asian countries because of scarcity of the conventional fuel. Most of the countries have greater number of rural population than its urban counterpart. The above discussed rural electrification programme in India through biomass gasifier will help to show a guideline for the countries like Cambodia, Laos, Burma and Vietnam especially. With the growing awareness of the importance of renewable energy sources and their potential role in decentralised energy generation along with their eco-friendly nature, the rate of growth of biomass energy systems is expected to accelerate in the future.

Biomass energy technologies still face a number of barriers in Asian countries, including subsidy for fossil fuels. The pace of commercialisation and deployment of these can be further facilitated through removal of a host of barriers these technologies face at present (Bhattacharya, 2002). By and large, research and development efforts of the Asian developing countries have not yet achieved any remarkable success. The relative successes of India appear to suggest that resource constraints have hindered RandD efforts in the smaller countries. A regional network would be useful in overcoming resource constraints faced by small countries in carrying out research and development efforts. Climate change-related developments are expected to create interest in and awareness of renewable energy sources, including biomass. Thus, renewable energy projects executed in developing countries can generate revenues though certified emissions reduction under the Clean Development Mechanism (CDM) of the Kyoto Protocol; this will make Renewable Energy projects more attractive. To meet the challenges of rising global concern regarding climate change Biomass gasifier power generation should not be an individual attempt rather a regional effort.

ACKNOWLEDGEMENT

Author gratefully acknowledges Professor Debesh Chakraborty (Dept. of Economics, Jadavpur University, Calcutta, India) for his valuable comments and suggestions on the work.

REFERENCES

Akshay Urja (2005) Renewable Energy, 1 (6), Nov-Dec

Bhattacharya S.C(2002) Biomass energy in Asia: a review of status, technologies and policies in Asia, Energy for Sustainable Development 1 Volume VI(31) September

CMIE(1994) Centre for Monitoring Indian Economy Report

Desi Power (2004) Advantages of decentralised biomass based power plants, Power to the People

Government of India, 2002: The Tenth Five Year Plan (2002-7),Vol II, Sectoral Policies And Programmes, Planning Commission, New Delhi

ITCOT (2005) "India: Biomass Energy for Rural India (PMU - BERI), BERI

Mukhopadhyay, K.(2004), An p5sessment of a Biomass Gasification based Power Plant in the Sunderbans, Biomass and Bioenergy, 27 (2004) :253 – 264

Ministry of Power, Rural Electrification Programme, Chapter 4, Annual report 2002-3, Government of India, New Delhi.

MNES (2004) Ministry of non conventional Energy Sources, Renewable Energy in India, Business opportunities, February.

MNES (2002) Renewable energy options in India

MNES (2005) Ministry of non conventional Energy Sources, Renewable Energy in India, February.

NSS (1992) National Sample Survey, Village level data on rural electrification.

Planning Commission (2005), Draft report of the expert committee on integrated energy policy, Government of India, New Delhi.

Rajvanshi A.k. (1995) Energy Self Sufficient Talukas - A Solution to National Energy Crisis, Economic and Political Weekly (EPW), Vol. XXX, pp. 3315-3319, 1995

Ravindranath, N. H. and Hall, D. O. (1995), Biomass, Energy, and Environment: A Developing Country Perspective from India, Oxford University Press.

Ravindranath (2004) Sustainable biomass power for rural India: Case study of biomass gasifier for village electrification, Current Science, Vol. 87, No. 7, 10 October 2004

Rehman, I.H. (2002) Non conventional energy and rural reconstruction, Yojana, January: 30-33

SEI (1999)--Stockholm environment Institute-- Renewable Energy for Development, September 1999, Vol. 12, No. 2/3 ---News letter of the Energy Program

Somashekhar, H.I. S. Dasappa and N.H. Ravindranath (2000), Rural bioenergy centres based on biomass gasifiers for decentralized power generation: case study of two villages in southern India, Energy for Sustainable Development, IV (3) October: 55-63

Thomas E.C (2002) Renewable Energy in India, Yojana, October,: 47-50

UNDP (2005) ---Equator Initiatives—Plant Power India (Series 5)

WEC (1993)—World Energy Council--Energy Information Centre Report

In: Progress in Biomass and Bioenergy Research
Editor: Steven F. Warnmer, pp. 141-154

Chapter 5

THE ENERGY BALANCE AND FUEL PROPERTIES OF BIODIESEL

Mustafa Acaroglu and Mahmut Ünaldı

Selçuk University, Technical Education Faculty Kampüs Konya Turkey

ABSTRACT

In this study energy balance and fuel properties of biodiesel has been calculated. Accordingly, the cost of 1 liter of oil is calculated 0.32 € after the income from the seed meal is deduced. Finally, the cost of per unit of biodiesel (1 liter) was calculated as 0.55 €, after deduction of the income provided by the sales of glycerin for use in soap and cosmetic industry.

The energy equivalent of total output was calculated 147605.50 MJ per hectare. The net energy gain (refined oil) was found as 15105.63 MJ per hectare (The net energy ratio 11.031) according to yield and inputs values.

The viscosity values of vegetable oils vary between 27.2 and 53.6 mm^2/s whereas those of vegetable oil methyl esters between 3.59 and 4.63 mm^2/s. The flash point values of vegetable oil methyl esters are highly lower than those of vegetable oils. The flash point values of vegetable oil methyl esters are highly lower than those of vegetable oils. An increase in density from 860 to 885 kg/m^3 for vegetable oil methyl esters or biodiesel increases the viscosity from 3.59 to 4.63 mm^2/s and the increases are highly regular. There is high regression between density and viscosity values vegetable oil methyl esters. The relationships between viscosity and flash point for vegetable oil methyl esters are irregular. An increase in density from 860 to 885 kg/m^3 for vegetable oil methyl esters increases the flash point from 401 to 453 K and the increases are slightly regular.

The LHV values of vegetable oils methyl ester vary between 35.74 and 39.16 MJ/kg.

Keywords: biodiesel, fuel, energy balance

1. INTRODUCTION

Biomass is a renewable resource so far as its production is continued in a sustainable way. Biomass conversion is a process to convert photosynthetic material into a more useful form. Fuels from biomass can take various forms such as solid, gas, and liquid. Liquid biomass (Vegetable oils) from rapeseed, safflower, soybean, palm oil, sunflower, and others can be used for diesel engine. Vegetable oils are used for food, as industrial raw materials, and for the generation of energy. Vegetable oils are a blend of free fatty acids (FFA's); monoglycerides, diglycerides, and triglycerides; phosphatides, lipoproteins, and glycolipids; waxes; terpenes, gums, and other less important compounds. Today's diesel engines require a clean-burning, stable fuel that performs well under a variety of operating conditions. Biodiesel is the only alternative fuel that can be used directly in any existing, unmodified diesel engine. Because it has similar properties to petroleum diesel fuel, biodiesel can be blended in any ratio with petroleum diesel fuel [1-3, 13].

Biodiesel is the name for a variety of ester-based oxygenated fuels made from vegetable oils or animal fats. Biodiesel is the name for a variety of ester-based oxygenated fuels made from vegetable oils or animal fats. The concept of using vegetable oil as a fuel dates back to 1895 when Dr. Rudolf Diesel developed the first diesel engine to run on vegetable oil. Biodiesel can be produced by several processes. Micro emulsification, pyrolysis transesterification and super critical method are the four different techniques used to production biodiesel. Vegetable oils or fats can be converted to fatty acids, which in turn are converted to esters. Oils or fats can also be converted to methyl or ethyl esters directly, using an acid or base to accelerate (catalyze) the transesterification reaction. Biodiesel is produced through a process known as transesterification, as shown in the equation below, where R1, R2, and R3 are long hydrocarbon chains, sometimes called fatty acid chains. There are only five chains that are most common in soybean oil and animal fats (others are present in small amounts) (Figure I) [4, 6, 8-11, 16-30].

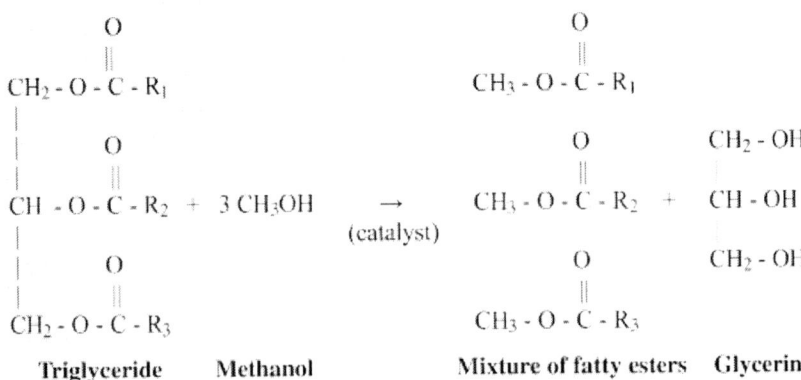

Figure I. Transesterification flow chart [1, 16].

Base catalyzation is preferred, because the reaction is quick and thorough. It also occurs at lower temperature and pressure than other processes, resulting in lower capital and operating costs for the biodiesel plant. The most common method of producing biodiesel is to reaction animal fat or vegetable oil with methanol in the presence of sodium hydroxide (a base, known as lye or caustic soda). This reaction is a base-catalyzed transesterification that produces methyl esters and glycerin. Methanol is preferred, because it is less expensive than ethanol. The majority of the alkyl esters produced today is done with the base catalyzed reaction because it is the most economic for several reasons:

- Low temperature (65 ^{0}C) and pressure (1.37*10^5 Pascal) processing.
- High conversion (98%) with minimal side reactions and reaction time.
- Direct conversion to methyl ester with no intermediate steps.

The use of biodiesel in a conventional diesel engine results in substantial reduction of unburned hydrocarbons, carbon monoxide, and particulate matter. Emissions of nitrogen oxides are either slightly increased depending on the duty cycle and testing methods. The use of biodiesel decreases the solid carbon fraction of particulate matter (since the oxygen in biodiesel enables more complete combustion to CO_2), eliminates the sulphate fraction (as there is no sulphur in the fuel), while the soluble, or hydrocarbon, fraction stays the same or is increased. Therefore, biodiesel works well with new technologies such as catalysts (which reduces the soluble fraction of diesel particulate but not the solid carbon fraction), particulate traps, and exhaust gas recirculation (potentially longer engine life due to less carbon). Biodiesel is less toxic than petroleum diesel and biodegrades as fast as dextrose. In addition, biodiesel has a flash point of over 125°C which makes it safer to store and handle than petroleum diesel fuel [15-28, 30, 31].

2. FUEL PROPERTIES OF BIODIESEL

Vegetable oils and their derivatives (especially methyl esters), commonly referred to as "biodiesel," are prominent candidates as alternative diesel fuels. Biodiesel is generally made of methyl esters of fatty acids produced by the transesterification reaction of triglycerides with methanol in the presence alkali as a catalyst [8]. Among the alcohols that can be used in the transesterification reaction are methanol, ethanol, propanol, butanol and amyl alcohol. Methanol and ethanol are used most frequently, ethanol is a preferred alcohol in the transesterification process compared to methanol because it is derived from agricultural products and is renewable and biologically less objectionable in the environment, however methanol because of its low cost and its physical and chemical advantages. The transesterification reaction can be catalyzed by alkalis [9, 10, 40, 61], acids [15], or enzymes [11, 18, 44, 46, 54].

The transesterification of triglycerides by supercritical methanol, ethanol, propanol and butanol, has proved to be the most promising process [1, 36, 37]. Supercritical methanol has a high potential for both transesterification of triglycerides and methyl esterification of free fatty acids to methyl esters for diesel fuel substitute. In the supercritical methanol transesterification method, the yield of conversion raises 95% for 10 minutes.

The catalyst (NaOH) is dissolved into methanol by vigorous stirring in a small reactor. The oil is transferred into the biodiesel reactor and then the catalyst/alcohol mixture is pumped into the oil. The final mixture is stirred vigorously for 1 hour at 338 K in ambient pressure. A successful transesterification reaction produces two liquid phases: ester and crude glycerin. Crude glycerin, the heavier liquid, will collect at the bottom after several hours of settling. Phase separation can be observed within 10 minutes and can be complete within 2 hours of settling. Complete settling can take as long as 20 hours. After settling is complete, water is added at the rate of 5.5 percent by volume of the methyl ester of oil and then stirred for 5 minutes and the glycerin is allowed to settle again. Washing the ester is a two-step process, which is carried out with extreme care. A water wash solution at the rate of 28 percent by volume of oil and 1 gram of tannic acid per liter of water is added to the ester and gently agitated. Air is carefully introduced into the aqueous layer while simultaneously stirring very gently. This process is continued until the ester layer becomes clear. After settling, the aqueous solution is drained and water alone is added at 28 percent by volume of oil for the final washing [1, 35, 36, 39, 41-43, 52-54, 57-61].

All the runs of supercritical methanol transesterification were performed in a 100-mL cylindrical. The sample was loaded from the bolt-hole into the autoclave, and the hole was plugged with a screw bolt after each run. In a typical run, the autoclave was charged with a given amount of vegetable oil (20-30 g) and liquid methanol (30-50 g) with changed molar ratios. The autoclave was supplied with heat from an external heater, and power was adjusted to give Zv approximate heating time of 15 min. The temperature of the reaction vessel was measured with an iron-constantan thermocouple and controlled at ±5 K for 30 min. Transesterification occurred during the heating period.

Eight different samples of biodiesel were used for viscosity, flash point and density measurements. A Redwood No. 1 viscosity meter with a measuring cup and a thermostat was used to measure the viscosity of all samples. The viscosity measurements were carried out at 313 K temperature. The temperatures were checked with a digital thermometer within the thermostat and the viscosity meter. At the beginning of each measurement a volume of 50 ml of the sample was filled into the measuring cup. We had to adjust shear rates for the different kinds of samples, because the viscosities are quite different and the viscosity meter has to be used within the correct measuring range. Flash point measurements were carried out using a Koehler mark apparatus.

Table 1. Fatty acid compositions of vegetable oil samples [1, 10]

Sample	16:0	16:1	18:0	18:1	18:2	18:3	Others
Cottonseed	28.7	0	0.9	13.0	57.4	0	0
Rapeseed	3.5	0	0.9	64.1	22.3	8.2	0
Safflowerseed	7.3	0	1.9	13.6	77.2	0	0
Safflowerseed	6.4	0.1	2.9	17.7	72.9	0	0
Palm	42.6	0.3	4.4	40.5	10.1	0.2	1.1
Soybean	13.9	0.3	2.1	23.2	56.2	4.3	0
Hazelnut kernel	4.9	0.2	2.6	83.6	8.5	0.2	0

**Table 2. Viscosity, density and flash point measurements
of ten vegetable oils [1, 17]**

Oil Source	Viscosity mm^2/s (at 311 K)	Density kg/m^3	Flash Point K
Corn	34.9	909.5	550
Cottonseed	33.5	914.8	509
Crambe	53.6	904.4	447
Linseed	27.2	923.6	514
Peanut	39.6	902.6	544
Rapeseed	37.0	911.5	519
Safflower	31.3	914.4	533
Sesame	35.5	913.3	533
Soybean	32.6	913.8	527
Sunflower	33.9	916.1	447

Table 3. Some fuel properties of six methyl ester biodiesels

Source	Viscosity cSt at 313.2 K	Density g/mL at 288.7 K	Cetane Number	Reference
Sunflower	4.6	0.880	49	[49]
Soybean	4.1	0.884	46	[53]
Palm	5.7	0.880	62	[49]
Penaut	4.9	0.876	54	[57]
Babassu	3.6	--	63	[57]
Tallow	4.1	0.877	58	[4]

**Table 4. Viscosity, density and flash point measurements
of eight oil methyl esters [3]**

Methyl ester	Viscosity mm^2/s (at 313 K)	Density kg/m^3 (at 288 K)	Flash point K
Cottonseed oil	3.69	880	437
Hazelnut kernel oil	3.59	860	401
Mustard oil	4.10	881	446
Palm oil	3.70	870	443
Rapeseed oil	4.63	885	428
Safflower oil	4.03	880	453
Soybean oil	4.08	885	447
Sunflower oil	4.22	880	443

Table 5. LHV (Low Heating Value) properties of methyl ester biodiesel[*]

Methyl ester	LHV (MJ/kg)	Ash (wt %)	Moisture (wt %)
Cottonseed oil	39.16		
Hazelnut kernel oil	38.70	0.013	0.75
Mustard oil	39.04	--	--
Palm oil	35.74	0.003	6.30
Rapeseed oil	38.24	0.035	1.30
Safflower oil	36.98	0.004	0.61
Sunflower oil	35.92	0.002	0.90

* Analysis time July 2005 (Acaroğlu)

3. BIODIESEL EMISSIONS COMPARED TO CONVENTIONAL DIESEL

Biodiesel is the first and only alternative fuel to have a complete evaluation of emission results and potential health effects submitted to the U.S. Environmental Protection Agency (EPA) under the Clean Air Act Section 211 (b). These programs include the most stringent emissions testing protocols ever required by EPA for certification of fuels or fuel additives in the US. The data gathered through these tests complete the most thorough inventory of the environmental and human health effects attributes that current technology will allow.

The overall ozone (smog) forming potential of biodiesel is less than diesel fuel. The ozone forming potential of the speciated hydrocarbon emissions was nearly 50 percent less than that measured for diesel fuel.

Sulfur emissions are essentially eliminated with pure biodiesel. The exhaust emissions of sulfur oxides and sulfates (major components of acid rain) from biodiesel were essentially eliminated compared to sulfur oxides and sulfates from diesel.

Criteria pollutants are reduced with biodiesel use. The use of biodiesel in an unmodified Cummins N14 diesel engine resulted in substantial reductions of unburned hydrocarbons, carbon monoxide, and particulate matter. Emissions of nitrogen oxides were slightly increased.

Carbon Monoxide - The exhaust emissions of carbon monoxide (a poisonous gas) from biodiesel were 47 percent lower than carbon monoxide emissions from diesel.

CO_2 emission index is defined as the CO_2 emission (%) divided by corresponding fuel consumption rate (in unit of g/h), the CO emission index is defined as CO emission (ppm) divided by the corresponding fuel consumption rate (in unit of g/h) and the NOx emission index is defined as NOx emission (ppm) divided by the corresponding fuel consumption rate. The catalytic converter reduced CO and HC emissions [1].

Benzene emissions increased with the amount of RME (rapeseed oil methylester) either with catalytic converter or without catalytic converter. This is remarkable because benzene is absent in RME. This indicates that the main source of benzene emissions may be a synthesis that occurs during combustion, rather than unburned fuel. However, the catalytic converter reduced the emissions by one third.

Particulate Matter - Breathing particulate has been shown to be a human health hazard. The exhaust emissions of particulate matter from biodiesel were 47 percent lower than overall particulate matter emissions from diesel,

Hydrocarbons - The exhaust emissions of total hydrocarbons (a contributing factor in the localized formation of smog and ozone) were 67 percent lower for biodiesel than diesel fuel.

Nitrogen Oxides(NOx) emissions from biodiesel increase or decrease depending on the engine family and testing procedures. NOx emissions (a contributing factor in the localized formation of smog and ozone) from pure (100%) biodiesel increased in this test by 10 percent. However, biodiesel's lack of sulfur allows the use of NOx control technologies that cannot be used with conventional diesel. So, biodiesel NOx emissions can be effectively managed and efficiently eliminated as a concern of the fuel's use [14, 26, 31, 33, 38, 49, 50, 59].

Biodiesel reduces the health risks associated with petroleum diesel. Biodiesel emissions showed decreased levels of PAH and nitrited PAH compounds which have been identified as potential cancer causing compounds. In the recent testing, PAH compounds were reduced by 75 to 85 percent, with the exception of benzo(a)anthracene, which was reduced by roughly 50 percent. Targeted nPAH compounds were also reduced dramatically with biodiesel fuel, with 2-nitrofluorene and 1 -nitropyrene reduced by 90 percent, and the rest of the nPAH compounds reduced to only trace levels.

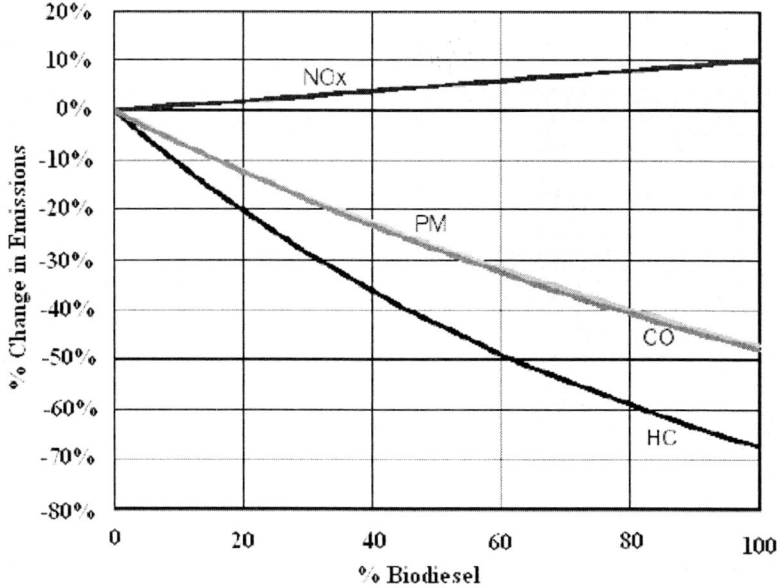

Figure 2. Average emission impacts of biodiesel fuels in CI engines [59].

4. THE ENERGY BALANCE OF SAFFLOWER OIL

Energy analysis, along with economic and environmental analyses, is an important tool to define the behavior of agricultural system. Energy analysis started as a relevant subject in agricultural production in the 1970's as a results of the dramatic increase of oil derivative prices. In an energy analysis of production systems it is necessary to consider the following steps ([1, 32, 47]:

- Set a limit in the process or system to be analyzed in such a way that all inputs and outputs, which pass that limit in a certain time interval, are evaluated.
- Assign energy requirements to all inputs
- Identify and quantify all outputs, establishing criteria for energy embodied in the main products and that corresponding to by-products.
- Relate output energy to total sequestered energy to obtain the energy ratio and the energy productivity.
- Apply energy analysis results

Safflower, Carthamus tinctorius, is among the oldest crops known to man. The Safflower (Carthamus tinctorius L.) is probably native to an area bounded by the Eastern Mediterranean and Persian Gulf. Believed to have originated in southern Asia and is known to have been cultivated in China, India, Persia and Egypt almost from prehistoric times. During middle Ages it was cultivated in Italy, France, and Spain, and soon after discovery of America, the Spanish took it to Mexico and then to Venezuela and Colombia. It was introduced into United States in 1925 from the Mediterranean region and is now grown in all parts west of 100th meridian. Safflower is commercially cultivated to a large extent in India, Mexico and the USA (Table 6, Table 7) [5, 12, 34, 48].

Table 6. Safflower in World (Country, Sown area, production and yield) [29]

Country	Sown Area (ha)	Production (ton)	Yield (kg/ha)
Indian	404100	226000	559
USA	79320	135160	1704
Ethiopia	72000	38000	528
Mexico	52758	52855	1002
Australia	35000	26000	743
Turkey	30	15	500
World	756055	577555	764

Table 7. Safflower in Turkey (year, sown area, production and yield)
(1995-2002), [29]

Years	Sown area (ha)	Production (ton)	Yield (kg/ha)
1995	134	125	930
1996	81	74	910
1997	74	65	880
1998	75	72	960
1999	50	50	1000
2000	30	18	600
2001	35	25	714
2002	30	15	500
Konya (2003-2004)	35	67.375	1925

Safflower is cultivated for the edible oil obtained from the seed. It contains a higher percentage of essential unsaturated fatty acids and a lower percentage of saturated fatty acids than other edible vegetable seed oils. *Safflower* oil lowers blood cholesterol levels and is used to treat heart diseases. The flowers have been the source of yellow and red dyes, largely replaced by synthetics, but still used in rouge. Seeds used for tumors, especially inflammatory tumors of the liver [19].

Flowers considered diaphoretic, emmenagogue, laxative, sedative, stimulant, in large doses laxative; used as a substitute or adulterant for saffron in treating measles, scarlatina, and other exanthematous diseases. Charred *safflower* oil used for rheumatism and sores; seeds, diuretic and tonic [7]. In China, prescribed as uterine astringent in dysmenorrhea. In Iran, the oil is used as a salve for sprains and rheumatism.

Safflower is propagated by seed. The seed is sown 2-3 cm deep, with planting distances 10 cm in the row and 30-60 cm between the rows. The desired crop density is 40-50 plants per m^2, and up to 70 plants per m^2 on light soil. The growing period is 200-250 days or 110-140 days, respectively [5, 48, 51].

No recommendations can be made for herbicide use, as this aspect of weed control is still being tested, but herbicides appropriate for sunflower crops are believed to be appropriate for *safflower* too [12]. The yields of seed can reach 1.1 – 1.7 t/ha/year [29]. With irrigation and good fertilization, yields 2.8-4.5 t/ha/year can be achieved, but the world average yield is 0.5 t/ha/year. The cultivars have 30-48 % oil in the fruit. The oil contains 73-79 % linoleic acid, depending on the cultivar [29, 45, 55, 56, 60].

In Turkey, today there are only three Safflower kinds. This study was used Dinçer variety (Table 8). Middle Anatolia region was selected and 35 ha Safflower was sowed in Konya. Sowing time was the first week of April (Table 9).

Table 8. The properties of *Safflower* variety crops [5]

Sort (variety)	Thorny	Flower Color	Plant height (cm)	Seed color	Oil Content (%)	Weight (gr/1000 seed)
Yenice	Not Thorny	Red	100-120	White	24-25	38-40
Dinçer	Not thorny	Orange	90-110	White	25-28	45-49
5-154	Thorny	Yellow	60-80	White	35-40	46-50

Table 9. The properties on sowing operations

Operation	Properties
Row distance	15 cm
Row upper	10 cm
Sowing deep	2.5-4.0 cm
Sowing (seeding) norm	20 kg/ha
Crop density	66 – 67 plant per m^2
Fertilizer (N)	120 kg/ha per year
Fertilizer (P_2O_5)	50 kg/ha per year
Fertilizer (K)	Not used

Harvesting is making with Combine harvester.

5. RESULTS AND DISCUSSION

In this study energy balance in Safflower production has been determined and calculated (Table 10 and Table 11). In making the energy balance, input and output values used in calculation are measured in field conditions and calculated using literature. Table 5 list the cultivation operations carried out, with the corresponding work hours and technical means and used with their respective consumption.

Table 10. Technical characteristic of machines and energy inputs *Safflower* (MJ/ha)

Agricultural Operation	(h/ha)	Tractors Energy	Equipment Energy	Fuel-lubricant Energy	Labor Energy	TOTAL
Primary Tillage Plough	2.14	73.36	14.35	604.53	4.00	696.24
Secondary tillage (Disk harrow)	0.84	28.91	10.16	238.33	1.58	278.97
Seeding	0.68	23.34	49.94	192.47	1.27	267.03
Fertilizer	0.22	7.55	1.96	62.18	0.41	72.11
Harvesting (Combine)	0.94		121.20	390.74	1.76	513.70
Total		133.16	197.62	1488.25	9.01	1828.04

Table 11. Fertilizer and Seed Energy Inputs

Fertilizer Energy	kg/ha	MJ/kg	MJ/ha
N	120	49.1	5892
P	40	17.78	711.2
		TOTAL	6603.2
Seed Energy	20	14	280

In harvesting, the Safflower seed yield was obtained as 1925 kg per hectare on an average and *Safflower* stalk was obtained as 6650 kg DM per hectare in the moisture of 15 % (18.08 MJ/kg). The average seed oil content was found 36 % (693 kg per hectare oil yield), (39.50 MJ/kg). (Table 12, Table 13).

Table 12. Total Energy Inputs of Safflower

Energy Input	MJ/ha	%
Fuel-oil	1488.25	12.13
Tractor Energy	133.16	1.09
Machinery Energy	197.62	1.61
Labour Energy	9.01	0.07
Fertilizer Energy	6603.20	53.83
Seed Energy	280.00	2.28
Industrial process (extraction)	2406.25	19,61
Industrial process (refining)	1150.38	9.38
TOTAL	**12267.87**	**100.00**

Table 13. Total Energy Outputs of Safflower

Energy Output	MJ/kg	MJ/ha
Safflower oil (Refined)	39.5	27373.5
Safflower stalk	18.08	120232
Safflower product residues	18.09	28871.6
Total		147605.5

Cost inputs in safflower oil biodiesel production has been calculated (table 14). In making the cost analysis, input and output values used in calculation are measured in process conditions and calculated using literature.

Table 14. Input and cost safflower biodiesel

Raw material	Cost per liter (€)	Conversion %	Quantity / liter biodiesel	Cost / liter biodiesel (€)
Oil	0.32	90	1.11	0.36
Methanol	0.250	-	0.20	0.05
NaOH	1.200	-	0.035	0.04
Various	-	-	-	0.01
Energy	0.100	-	-	0.10
Biodiesel cost	-	-	-	0.55

The energy equivalent of total output was calculated 147605.50 MJ per hectare. The net energy gain (refined oil) was found as 15105.63 MJ/ha (The net energy ratio 11.031) according to yield and inputs values. Net energy gain of Safflower seed is higher then other oil seeds in the Turkey. In addition to these, Safflower crops will have a great importance in Turkey as oil resource and biomass energy source as it solves environmental problems, its inputs are low, and it can be grown in arid zones.

6. REFERENCES

[1]   Acaroğlu, M. 2003. Renewable Energy Sources, Atlas Yayın and Dağıtım, Istanbul, (Turkish Book)

[2]   Acaroğlu, M., Gezer, I., 2004, Renewable Energy Sources and Future of Biomass Energy in Turkey, pp. 413-416, 2nd World Conference on Biomass for Energy, Industry and Climate Protection, 10-14 May 2004, Rome, Italy

[3]   Acaroğlu, M., Demirbaş, A., 2005, Relationships between Viscosity and Density Measurements of Biodiesel Fuels, *Energy Sources* (in press).

[4]   Ali Y, Hanna MA, Cuppett SL. 1995. Fuel properties of tallow and soybean oil esters. *JAOCS* 1995; 72:1557-1564.

[5]   Babaoğlu, M., 2004 Dünya'da ve Türkiye'de Aspir Bitkisinin Tarihçesi, Kullanım Alanları ve Önemi (Turkish), Trakya Tarımsal Araştırma Enstitüsü Edirne

[6]   Bala, B. K. 2005. Studies on biodiesels from transformation of vegetable oils for diesel engines, *Energy Edu. Sci. Technol.* 15:1-43.

[7]   C.S.I.R. (Council of Scientific and Industrial Research), 1948-1976, *The wealth of India.* 11 vols, New Delhi

[8]   Clark,S. J., L. Wagner, M. D. Schrock, and P. G. Pinnaar. 1984. Methyl and ethyl esters as renewable fuels for diesel engines. *JAOCS* 61:1632–1638.

[9]   Demirbas A. Biodiesel from vegetable oils via transesterification in supercritical methanol, *Energy Convers. Mgmt.* 2002; 43:2349–56.

[10]   Demirbas A. Biodiesel fuels from vegetable oils via catalytic and non-catalytic supercritical alcohol transesterifications and other methods: a survey. *Energy Convers Management* 2003;44:2093-2109.

[11]   Du W, Xu Y, Liu D, Zeng J. Comparative study on lipase-catalyzed transformation of soybean oil for biodiesel production with different acyl acceptors, *J. Molecular Catal. B: Enzymatic.* 2004;30:125–129

[12]   Duke, J.A., 1983. *Handbook of Energy Crops* (unpublished).

[13]   El Bassam, N., 1998. Energy Plant Species, Their use and impact on environment, JamesandJames (Science Publishers), 320 p., *ISBN 1* 873936 75 3

[14]   EPA, 2002, A Comprehensive Analysis of Biodiesel Impacts on Exhaust Emissions, *Draft Technical Report,* EPA420-P-02-001 October 2002

[15]   Furuta S, Matsuhashi H, Arata K. Biodiesel fuel production with solid superacid catalysis in fixed bed reactor under atmospheric pressure. *Catal. Commun.* 2004; 5:721–723.

[16]   Gerpen, V., 2005, Biodiesel processing and production, Fuel Processing Technology 86 (2005) 1097–1107

[17]   Goering, E., W. Schwab, J. Daugherty, H. Pryde, and J. Heakin, 1982, Fuel properties of eleven vegetable oils. *Trans ASAE* 25: 1472–1483.

[18]   Hama S, Yamaji H, Kaieda M, Oda M, Kondo A, Fukuda H. Effect of fatty acid membrane composition on whole-cell biocatalysts for biodiesel-fuel production. *Biochem. Eng. J.* 2004;21:155–60

[19]   Hartwell, J.L. 1967-1971. Plants used against cancer. A survey. Lloydia 30-34.

[20]   http://journeytoforever.org/biodiesel.html

[21]   http://ww2.green-trust.org:8383/biodiesel2.htm

[22] http://www.biodiesel.org

[23] http://www.biodiesel.org/pdf_files/fuelfactsheets/emissions.pdf

[24] http://www.canentec.com/whatisbiodiesel.html

[25] http://www.distributiondrive.com/Article16.html

[26] http://www.epa.gov/otaq/models/biodsl.htm

[27] http://www.fact-index.com/b/bi/biodiesel.html#History

[28] http://www.fact-index.com/m/ma/main_page.html

[29] http://www.fao.org

[30] http://www.liquid-biofuels.com/pub.htm

[31] http://www.soypower.net/BiodieselPDF/BiodieselEmissions.pdf

[32] Intosh, C. S., Withers, R. V., Smith, S. M., 1982. The Economics of On-Farm Production and Use of Vegetable Oils for Fuel, Paper No.177, Proceedings of the International Conference on Plant and Vegetable Oils as Fuels, Holiday Inn Fargo North Dakota.

[33] Kaltschmitt, M., Reinhardt, G.A., 1997. Nachwachsende Energieträger - Grundlagen, Verfahren, ökologische Bilanzierung. Vieweg-Verlag. Braunschweig.

[34] Keys, J.D. 1976, Chinese herbs, their botany, chemistry, and pharmacodynamics, Chas. E. Tuttle Co., Tokyo

[35] Krawczyk, T., "Biodiesel – Alternative fuel makes inroads but hurdles remain," *INFORM* vol. 7, no. 8(1996), pp.801-815, American Oil Chemist Society

[36] Kusdiana D, Saka S. Effects of water on biodiesel fuel production by supercritical methanol treatment. *Biores.Technol.* 2004; 91: 289-295.

[37] Kusdiana D, Saka S. Kinetics of transesterification in rapeseed oil to biodiesel fuels as treated in supercritical methanol. *Fuel* 2001; 80: 693–698.

[38] Lin, C.Y., Lin, H.A, 2005, Diesel engine performance and emission characteristics of biodiesel produced by the peroxidation process, *Fuel* xx (2005) 1-8.

[39] Lyell, K., 2003. Design of a Methanol Extraction Process for Bio-Diesel Production, Fall 2003, *Austin Bio-Fuels,* LLC, Texas.

[40] Ma F, Hanna MA. Biodiesel production: a review. Biores Technol 1999, 70:1-15.

[41] Ma, F., "Biodiesel fuel: The transesterification of beef tallow." PhD dissertation, *Biological Systems Engineering*, University of Nebraska-Lincoln (1998)

[42] Ma, F., Clements, L.D., Hanna, M.A, 1998. Biodiesel fuel form animal fat. Ancillary studies on transesterification of beef tallow, *Ind. Eng. Chem. Res.* vol.37 (1998b), pp. 3768-3771.

[43] Ma, F., Hanna, M.A., Biodiesel production: a review, *Biosource Technology* vol.7 (1999), pp.1-15.

[44] Noureddini H, Gao X, Philkana RS. Immobilized Pseudomonas cepacia lipase for biodiesel fuel production from soybean oil, *Biores. Technol.* 2005; 96:769–777.

[45] O'Brien, R., 1998, Fats and Oils, Formulating and Processing for Applications, 667 p., Technomic Publishing AG, Basel, Switzerland

[46] Oda M, Kaieda M, Hama S, Yamaji H, Kondo A, Izumoto E, Fukuda H. Facilitatory effect of immobilized lipase-producing *Rhizopus oryzae* cells on acyl migration in biodiesel-fuel production. *Biochem. Eng. J.* 2004; 23:45–51.

[47] Ortiz-Canavate, J., Hernanz, J.L., 1999. Energy for Biological Systems, Energy Analysis and Saving, CIGR Handbook of Agricultural Engineering, Energy and Biomass Engineering, pp.13-42, Published by ASAE, USA.

[48] Ozturk, O, 2004. Aspir Tarımının Önemi ve Orta Anadolu Şartlarında Yetiştirme İmkanları, Konya Ticaret Borsası, April 2004, Year: 7, N. 17, pp. 54-60 (Turkish), Konya, Turkey.

[49] Pischinger GM, Falcon AM, Siekmann RW, Fernandes FR. Methylesters of plant oils as diesels fuels, either straight or in blends. Vegetable Oil Fuels, ASAE Publication 4-82, *Amer. Soc. Agric. Engrs. St. Joseph, MI, USA,* 1982.

[50] Pryor, R. W., M. A. Hanna, J. L. Schinstock, and L. L. Bashford. 1982. Soybean oil fuel in a small diesel engine. *Trans ASAE* 26:333-338.

[51] Raghu, J.S. and Sharma, S.R. 1978, Response to irrigation and fertility levels of *safflower*. Indian *J. Agron.* 23(2):93-97.

[52] Riva, G., Sissot, F., 1999. Vegetable Oils and Their Esters (biodiesel), CIGR Handbook of Agricultural Engineering, Energy and Biomass Engineering, pp. 164-201, ASAE, USA.

[53] Schwab AW, Bagby MO, Freedman B. Preparation and properties of diesel fuels from vegetable oils. *Fuel* 1987;66:1372–1378.

[54] Shieh C-J, Liao H-F, Lee C-C. Optimization of lipase-catalyzed biodiesel by response surface methodology. *Bioresource Technology* 2003;88:103–106.

[55] Smith, J.R., 1985. "Safflower: Due for a Rebound", *J. Am. Oil Chem. Soc.,* 62(9):1286-1291.

[56] Sonntag, N.O.V., 1979. "Composition and Characteristics of Individual Fats and Oils" in Bailey's Industrial Oil and Fat Products, Vol. 1, 4th Edition, D. Swern, ed. New York, NY: A Wiley- Interscience Publication, pp. 398-403.

[57] Srivastava A, Prasad R. Triglycerides-based diesel fuels. *Renew. Sustain. Energy Rev.* 2000;4:111–133

[58] Tickell, J., 2000.From the Fryer to the Fuel Tank the Complete Guide to Using Vegetable Oil as an Alternative Fuel. *Tickell Energy Consulting,* USA

[59] Tyson, K.S., 2004, *Biodiesel Handling and Use Guidelines*, DOE/GO 102004-1999 Revised November 2004; U.S. Department of Energy.

[60] Weiss, T.J., 1983. "Commercial Oil Sources", in Food Oil and Their Uses, Second Edition

[61] Zhang Y, Dub MA, McLean DD, Kates M. Biodiesel production from waste cooking oil: 2. Economic assessment and sensitivity analysis. *Biores. Technol.* 2003; 90:229–240.

In: Progress in Biomass and Bioenergy Research
Editor: Steven F. Warnmer, pp. 155-175

ISBN: 1-60021-328-6
© 2007 Nova Science Publishers, Inc.

Chapter 6

AN EXPERIMENTAL STUDY ON PERFORMANCE AND EXHAUST EMISSIONS OF A DIESEL ENGINE FUELLED WITH VARIOUS BIODIESELS

*Nazim Usta**

Pamukkale University, Mechanical Engineering Department
Camlık 20017 Denizli, Turkey

ABSTRACT

Instability and increases in prices of petroleum-based fuels, gradual depletion of world petroleum reserves and increases in environmental pollution caused by exhaust emissions speed up research on renewable alternative fuels.

Vegetable oils have been considered as renewable alternative fuels in compression ignition engines for a long time. However, they have not been widely used as fuels in the engines due to some technical and economical drawbacks. Some properties of vegetable oils such as high viscosity, lower volatility and lower heat content result in technical problems in direct using of vegetable oils in short and long term applications. From economical point of view, the main problem is that vegetable oils have been more expensive than petroleum Diesel fuel.

There are various ongoing studies on solving these problems to be able to use vegetable oils in Diesel engines. Different methods such as preheating oils, blending or dilution with other fuels, thermal cracking/pyrolysis and transesterification have been developed. Among these techniques, transesterification appears to be the most promising one. It is a chemical process converting vegetable oils to alcohol ester of oil named as biodiesel. In general, biodiesel-Diesel fuel No.2 blend can be used as a fuel in Diesel engines without modification. Specifications of biodiesel mainly depend on oil, transesterification process, type and amount of alcohol, type and amount of catalysis, reaction time and temperature.

* Nazim Usta; Pamukkale University; Mechanical Engineering Department; Camlık 20017 Denizli; Turkey; Tel: +90 258 2125532; Fax: +90 258 2125538; E-mail: n_usta@pamukkale.edu.tr ; usta_n@yahoo.com

Biodiesel can be produced from different kinds of vegetable oils. Since prices of edible vegetable oils are higher than that of Diesel fuel No. 2, waste vegetable oils and non-edible crude vegetable oils are mostly preferred as potential low priced biodiesel sources. It is also possible to use soapstock, a by-product of edible oil production, for cheap biodiesel production.

In this study, various biodiesels were produced from raw vegetable oils (rapeseed oil, soybean oil, cotton seed oil, palm oil and tobacco seed oil), waste sunflower vegetable oils and hazelnut oil soap stock-waste sunflower vegetable oil, and their specifications were compared with each other. The biodiesel (20% in volume) - Diesel fuel No.2 (80% in volume) blends were tested in a four cycle, four cylinder turbocharged indirect injection Diesel engine. The effects of biodiesel addition to Diesel fuel No.2 on the performance and emissions of the engine were investigated at full load. Experimental results showed that the biodiesels can be partially substituted for Diesel fuel No.2 at most operating conditions in terms of performance parameters and emissions without any engine modification and preheating of the blends.

Keywords: Biodiesel, diesel engine, performance, emission.

NOMENCLATURE

C	cotton seed oil methyl ester
D	Diesel fuel No.2
P	palm oil methyl ester
R	rape seed oil methyl ester
S	soybean oil methyl ester
SOW	mixture (waste sunflower oil in 50%–hazelnut kernel soap stock in 50%) methyl ester
T	tobacco seed oil methyl ester
W	waste sunflower oil methyl ester

1. INTRODUCTION

Increases in prices of petroleum-based fuels, the environmental pollution due to exhaust emissions from these fuels, the gradual depletion of world petroleum reserves and economical and political instabilities in countries exporting petroleum have encouraged studies to search for alternative renewable fuels.

Vegetable oils have been considered as alternative renewable fuels for compression ignition (CI) engines which are mainly used in transport sector. Vegetable oils are non-toxic, biodegradable, and have low emission profiles (Williamson and Badr, 1998; Ma and Hanna, 1999; Srivastava and Prasad, 2000; Kalam et al. 2003; Kalligeros et al. 2003). Use of vegetable oils in CI engines is known since the invention of CI engine by Rudolph Diesel. He used peanut oil in the engine. However, vegetable oils have not been widely and effectively used in CI engines due to some drawbacks such as higher viscosity, lower volatility and lower heat content (Ergeneman et al., 1997a; Altin et al., 2001; Nwafor, 2003). Especially the higher viscosity which causes poor atomization and results in incomplete combustion is an important disadvantage of vegetable oils. It may also lead to formation of injector deposits, ring sticking, development of gumming, as well as incompatibility with lubricating oils in long term operations (Williamson and Badr, 1998; Karaosmanoğlu et al. 2000). In addition, prices of vegetable oils have been higher than that of diesel fuel No.2 and this is an important economical drawback related to use of vegetable oil in CI engines.

Researchers have been tried to develop different techniques such as preheating oils, blending or dilution with other fuels, thermal cracking/pyrolysis and transesterification to use vegetable oils in CI engines effectively and efficiently (Williamson and Badr, 1998; Ma and Hanna, 1999; Karaosmanoglu, 1999; Agarwal and Das, 2001; Demirbaş, 2003). Among these techniques, transesterification which is a chemical process of converting vegetable oil into methyl or ethyl ester of vegetable oil has been widely preferred and used successfully (Kalligeros et al. 2003; Agarwal and Das, 2001; Demirbaş, 2002; Kumar et al., 2003). In general, the methyl or ethyl ester of vegetables is called as biodiesel. The transesterification process has been modified depending on type and condition of the vegetable oil. The detailed reviews about the process are available in the literature (Ma and Hanna, 1999; Demirbaş, 2002; Agarwal and Das, 2001; Karaosmanoglu, 1999; Demirbaş, 2003; Ozaktas et al. 1997).

There are different kinds of sources for biodiesel production. Although virgin edible vegetable oils are mainly used in food sectors, rapeseed oil, soybean and palm oil have been mainly used for biodiesel production (Crabbe et al., 2001; Kumar et al. 2003). There are some studies related to non-edible vegetable oils such as tobacco seed oil (Usta, 2005). To the author's best knowledge, tobacco seed oil was used first time by the author for biodiesel production and usage as a fuel in a diesel engine. There are also some other studies on non-edible vegetable oils (Srivastava and Prasad, 2000; Agarwal and Das, 2001; Pramanik, 2003). The extensive usage of biodiesel may results in the finding of new sources. Waste cooking oil is an important cheap source for biodiesel production (Gonzalez Gomez et al. 2000; Dorado et al., 2003; Ozaktas, 2000; Alcantara et al., 2000; Al-Widyan and Al-Shyoukh, 2002; Tomasevic and Siler-Marinkovic, 2003). In addition, soapstocks have been considered to use in biodiesel production (Haas et al., 2000; Haas et al., 2001; Haas and Foglia, 2002; Haas et al., 2003; Graboski et al., 2003; Usta et al., 2005). Soapstocks are also cheap sources for biodiesel production.

Biodiesel can be used in different proportions in CI engines without any modification. In the US, the most commonly used blend is 20 % (in volume) biodiesel in diesel fuel No.2, which is called as a B20 blend. However, in Europe the most commonly used blend is 5 % (in volume) in the diesel fuel (Piazza and Fogila, 2001).

There is an important compositional difference between biodiesels and the diesel fuel. Biodiesels contain approximately 10–12% oxygen in weight basis. This leads to reduction in the energy content of the fuel resulting in lower engine torque and power (Altin et al., 2001; Nwafor et al., 2000; Bari et al. 2002; Antolín et al., 2002). However, the oxygen in the fuel helps to reduce exhaust emissions such as smoke, CO and HC mainly due to the effect of complete combustion (Kalam et al., 2003; Kalligeros et al., 2003; Gonzalez Gomez et al. 2000; Ozaktas et al., 1997; Ergeneman et al., 1997b; Marshall et al., 1995; Chang and Van Gerpen, 1997; Monyem and Van Gerpen, 2001; Graboski and McCormick, 1998; Curran et al., 2001; Kitamura et al., 2001a; Kitamura et al., 2001b).

Since vegetable oils includes very little sulphur compared to the diesel fuel No.2, some reduction in SO_2 emission is obtained depending on the proportion of biodiesel in the fuel (Dorado et al., 2003). The main disadvantage of biodiesel on emissions is related to NO_x. Although NO_x emissions mainly depend on the engine fuelling system, engine type and engine loading, in general biodiesel usage increases NO_x emissions due to oxygen content of the fuel and higher temperatures of combustion chamber (Gonzalez Gomez et al., 2000).

In this study, biodiesels produced from seven different sources (rapeseed oil, soybean oil, cotton seed oil, palm oil, tobacco seed oil, waste vegetable oils and hazelnut kernel oil soap stock-waste vegetable oil mixture) were blended with diesel fuel No.2 in 20 % (in volume). The blends were tested in a four cycle, four cylinder turbocharged indirect injection diesel engine. The effects of biodiesel addition in 20% to Diesel No. 2 on the performance and emissions of the engine were investigated at full load. All tests were performed without any modification on the engine.

2. MATERIALS AND METHODS

2.1. Raw Materials

In this section, the seven different sources, which are important sources for biodiesel productions worldwide and used in this study, are described briefly.

Rapeseed is one of the most important sources of vegetable oil in the world. Natural rapeseed oil contains erucic acid which is a kind of toxic material to humans in large doses. An edible form of rapeseed (canola) which has low erucic acid content was developed in Canada and then the rapeseed production has increased. In Europe, low-erucic rapeseed oil is the major source for biodiesel production (Piazza and Fogila, 2001).

Soybean oil is the world's most widely used edible oil and it is a very healthy food ingredient. It is not only used in food products but is also used to produce some non-food products like biodiesel, inks, plasticizers, crayons, paints and soy candles. It is the primary oil of biodiesel production in the United States (Piazza and Fogila, 2001).

Cottonseed oil is edible oil and it is mainly used in food sector. However, it is less favored than sunflower, corn and soybean oils in foods. Meanwhile the cotton seed oil which is not used in the food sector may be utilized oil in biodiesel production, especially in certain countries where cotton is widely produced (Karaosmanoğlu et al.,1999).

Palm oil is an edible vegetable oil which is extracted from the fruit of oil palm tree. In addition it can be used in non-food products. It is also likely to be used in biodiesel production. The high oil yield per hectare area has made it the main source of vegetable oil for many tropical countries. Main palm oil producers and exporters in the world are Malaysia and Indonesia.

Tobacco plant is a beautiful plant with large oval leaves, pink flowers and green capsules containing numerous very small seeds (U.S. Department of Agriculture, 2004). The seed has a strong shell, therefore it can resistant to high humidity and can be stored in dry conditions at ordinary temperatures. The oil content of seeds changes from 36% to 41% by weight (Giannelos et al., 2002). The rest of the seed consists of protein, carbohydrate, inorganic material and crude fiber. The plant is grown in 119 countries in the world mainly for leaves which are commercial products and used in the production of cigarettes in the tobacco processing industries. Since the oil which can be extracted from tobacco seeds is a non-edible oil and not used in other applications, small amount of tobacco seeds is collected from fields for next year production. But most of them are left unused in fields. Tobacco harvesting area and leaves production of different countries are available in (U.S. Department of Agriculture, 2004). However, there is no statistical information on potential tobacco seeds in the literature to the best knowledge of the author. Meanwhile, the seed and oil potential may be estimated using the harvesting area information. Tobacco seed oil was used first time by the author for the biodiesel production. Since the tobacco seed oil is not available in markets it was required to extract the oil from the seeds. The detailed information can be found in a paper of the author (Usta, 2005a).

Waste cooking oil is a cheap source for biodiesel production. However the operational cost of biodiesel production from this oil is higher than that from virgin oil due to the requirement of filtration and water removal. In addition, the properties of waste cooking oils vary depending on the raw oil and the cooking application. One of the important properties of waste cooking oil is the free fatty acid content. If the free fatty acid content of the oil is higher than %2, it is required to use acid catalyser or acid/base catalysers in biodiesel production.

Soapstock is a by-product of edible vegetable oil production and also a cheap source for biodiesel production. In general it is used in soap industry. There are limited numbers of studies on soapstock as a biodiesel source in the literature as mentioned above. Since soapstock contains large amounts of free fatty acids (45–50%), it cannot be effectively converted to biodiesel using only an alkaline catalyst. It is required to reduce the free fatty acids of the feedstock using an acid catalysed pre-treatment to esterify the free acids before transesterifying the triglycerides with an alkaline catalyst to complete the reaction.

The fatty acid composition of vegetable oils is the one of the important parameter which affects specification of their methyl esters. Therefore the fatty acid compositions of the oils are given in Table 1.

Table 1. Fatty Acid Compositions of the Vegetable Oils (% by weight)

Fatty acid	Tobacco seed [*]	Soybean [**]	Rapeseed [**]	Cotton seed [**]	Palm [**]	Sunflower seed [**]	Hazelnut kernel [**]
Palmitic (16.0)	10.96	13.9	3.5	28.7	42.6	6.4	4.9
Palmitoleic (16:1)	0.2	0.3	0.0	0.0	0.3	0.1	0.2
Stearic (18:0)	3.34	2.1	0.9	0.9	4.4	2.9	2.6
Oleic (18:1)	14.54	23.2	64.1	13.0	40.5	17.7	83.6
Linoleic (18:2)	69.49	56.2	22.3	57.4	10.1	72.9	8.5
Linolenic (18:3)	0.69	4.3	8.2	0.0	0.2	0.0	0.2
Others	0.78	0.0	1.0	0.0	1.9	0.0	0.0

* Giannelos et al. (2002)
**Demirbaş (2003)

2.2. Biodiesel Production

There are different kinds of biodiesel production techniques. Researchers try to increase yield and quality of biodiesels. The quality of a biodiesel is checked with standards, mainly ASTM 6751 and EN14214.

The general procedure applied in this study for the virgin oils (rapeseed, tobacco seed, soybean and palm oils) can be summarised as follows: the oils were converted into methyl esters by a transesterification process in which the triglycerides of the oils react with methyl alcohol in the presence of a base catalyst to produce glycerol and fatty acid esters. 6/1 molar ratio of alcohol to triglycerides, twice of stoichiometric ratio, was used to get the largest ester yield (Al-Widyan and Al-Shyoukh, 2002; Tomasevic and Siler-Marinkovic, 2003; Darnoko and Cheryan, 2000). NaOH as a catalyst was used due to its high activity (Vicente et al. 1998) and the amount of NaOH was determined using a titration process as advised in (Tickell and Tickell, 1999). Since methyl alcohol was used in the process, 55°C which is below the boiling temperature of methanol was chosen for the reaction temperature. NaOH is dissolved in methanol to produce the methoxide and then it is poured into the oil which is heated to 55°C previously. The solution was stirred for 1.5 h holding the temperature at 55°C, then the heater is turned off and the stirring is continued for 1.5h without heating. The mixture is allowed to form two layers overnight. The upper layer was the ester while the bottom layer was glycerine. The glycerine was taken out. The remaining ester was washed with pure water three times. The water is settled at bottom and it was removed. At the end of the process the ester was heated to 100°C to remove unused methanol and water from the oil left in the ester.

The waste cooking oil also was transesterified as the process explained as above. Because the free fatty acid content of the oil is lower than 2%. However, it is required to be careful, if the free fatty acid content of the oil is higher than 2%, the alkaly catalyser is not advised to prevent soap production. In that case, it is advised to use acid/base catalyser process as explained below.

Soap stocks have very high free fatty acid (FFA) (45–50%). When a base catalyser is directly applied to the soap stocks, the high free fatty acid content causes high soap formation (Canakci and Van Gerpen, 2001). Therefore they can not be esterified with alkali catalysers. It is required acid or acid/base catalyser. Haas et al. (2000) reported that fatty acid methyl esters (FAME) from soybean soapstock can be produced in a two stage process that involves alkaline hydrolysis of all lipid linked fatty acid ester bonds and acid catalysed esterification of the resulting fatty acid sodium salts. In addition Haas et al. (2003) reported that FAME can be produced from soapstock using only acid catalysed esterification. They found the maximum esterification occurred at 65°C and 26 h reaction at a molar ratio of total fatty acid (FA)/methanol/sulphuric acid of 1:15:1.5. In this study hazelnut soapstock and waste sunflower oil mixture was used as a source in approximately equal volume proportions. Although the waste sunflower oil reduced the FFA content of the mixture, the mixture was still not suitable for base catalyser. In this case the free fatty acid content of the mixture was reduced using an acid catalysed pre-treatment at 35°C to esterify the free acids before transesterifying the triglycerides with an alkaline catalyst to complete the reaction at 55° C.

The acid/base catalyser process is summarised as follows: After the removing particles and water from the mixture, methanol (8% of the mixture in volume) was added to the mixture which is at 35°C, and it was stirred for 5 min. Then one millilitre of 95% sulphuric acid was added to the mixture-methanol. The stirring was continued for 1 h with heating keeping the temperature constant at 35°C and for 1 h without any heating. The mixture was left overnight. In the second stage, 3.5 g NaOH per litre of the mixture was dissolved in methanol (12% of the hazelnut soapstock-waste sunflower oil mixture) to produce methoxide. Initially half of the methoxide was poured into the mixture and mixed for 5 minutes. Then the mixture was heated to 55°C, and the rest of the methoxide was added to the heated mixture. The stirring was continued for 90 minutes. The mixture was allowed to form two layers overnight, sometimes it may take more time. The bottom layer was glycerine, while the upper layer was the ester. The rest of the process is similar to the process mentioned as above.

2.3. Experimental Apparatus for Engine Tests and Exhaust Emissions

The fuel testing system consists of a Cussons-P8651 type engine test bed, a four cylinder four stroke turbocharged indirect injection diesel engine, a gas analyser and a smokemeter. The schematic of the system is shown in Fig. 1. The specifications of the diesel engine are given in Table 2. The engine test bed consists of a hydraulic dynamometer, measurement instruments and a control panel. The water cooled hydraulic dynamometer is rated for 112 kW power absorption at 9000 rpm. A strain gauge load sensor which was calibrated by using standard weights just before the experiments was used to measure the load on the dynamometer. The speed of the engine was measured using inductive pickup speed sensor calibrated by an optical tachometer. The air flowrate was measured by means of an air box, a venturi meter and a manometer. The fuel flowrate was measured with a burette with 50 and 100 ml volumes and a stopwatch. A mechanical actuator was used to adjust different loads. A data logger with K type thermocouples was used to measure air inlet, fuel, engine coolant inlet-outlet, lubricating oil, exhaust gas temperatures. A second fuel tank was fixed to the system for alternative fuels.

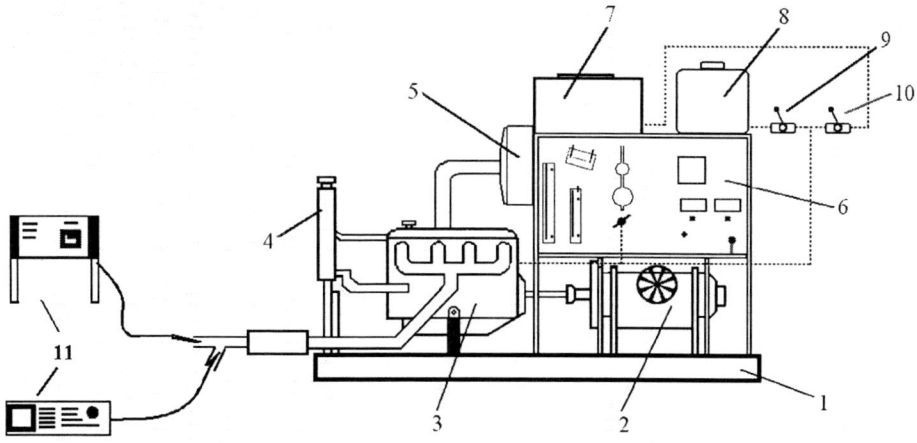

Figure 1. Experimental rig (1-Engine chassis, 2- Hydrokinetic dynamometer, 3-Engine tank, 4-Engine cooling unit, 5-Air tank 6- Control unit 7-Main fuel tank, 8-Alternative fuel tank, 9-Biodiesel control valve, 10-Diesel control valve, 11-Exhaust gas analyser and smokemeter).

Table 2. Specifications of the diesel engine

Type	Ford XLD 418T Turbocharged 4 Stroke, Water Cooled, IDI
Number of Cylinder	4
Stroke	82.0 mm
Bore	82.5 mm
Compression Ratio	21.5:1
Displacement	1.753 litres
Maximum Torque	152 Nm at 2200rpm
Maximum Power	55kW at 4500 rpm
Type of Injection Pump	Rotary Distributor

A Testo 350 M/XL gas analyzer was used to measure CO, NO, NO_2, H_2S, O_2 and SO_2 emissions. The smoke was measured using a Bosch BEA 170 smokemeter. Table 3 shows the accuracies of the measurements and the uncertainties in the calculated results.

**Table 3. The accuracies of the measurements and the uncertainities
in the calculated results**

Measurements Accuracy	
Load	∓ 2 N
Speed	∓ 2 rpm
Time	∓ 0.5 %
Temperatures	∓ 1 °C
CO	∓ 20 ppm
K	∓ 0.1 %
NO_x	∓ 20 ppm
SO_2	∓ 20 ppm
Dynamic Viscosity	∓ 1 %
Specific gravity	∓ 1 %
Calculated Results Uncertainty	
Kinematic Viscosity	∓ 1.4 %
Power	∓ 2 %
bsfc	∓ 2.3 %
Thermal Efficiency	∓ 2.5 %

Diesel fuel No.2 and blends containing 20% the biodiesels by volume were tested in the engine at full load. The engine was run approximately 30 min to warm up and then the engine speed was increased to 3000 rpm. The measurements were done at five different engine speeds, namely 3000, 2500, 2200, 2000 and 1500 rpm. At each speed, the engine was run approximately four minutes and then the measurement parameters were recorded at fifth minute. For biodiesel blends experiments, the diesel fuel valve was shut down and the blend valve was opened to run the engine with the biodiesel blends.

In this study, abbreviations for diesel fuel No.2 and the biodiesels are given in the nomenclature. In addition numbers just after the abbreviations show the percentage of the fuel in the blends. For example, the legend D100 represents 100% the diesel fuel No.2, while R20 indicates a blend containing 80% the diesel fuel No.2 and 20% rape seed oil methyl ester.

3. EXPERIMENTAL RESULTS AND DISCUSSIONS

3.1. Comparison of Biodiesel's Specifications

In general, it is known that specifications of a biodiesel depend on the oil and the transesterification process. After production of a biodiesel, the specifications should be determined and checked whether they are within the limits of standards, mainly ASTM D6751 and EN14214. Some specifications such as density, viscosity and cetane number mainly depend on the fatty acid composition of the oil. Therefore if these specifications are out of the limits, it is required to use some additional improvers before using the biodiesel. However, some other specifications such as water content, monoglyceride, diglyceride, triglyceride content, free glycerol and total glycerol mainly depend on the transesterification process. These specifications can be improved by modifications in the process. Different biodiesels can be produced from the same oil using different processes.

In most cases, the specifications of biodiesels produced and examined in this study are within the limits of EN14214 standard, except SOW100. Since the biodiesels were tested in a diesel engine as a blend (20% in volume), the study was focused on the performance and emission tests rather than the details of specifications. However, the densities and viscosities of the biodiesels are compared with D100. The changes of viscosities with temperature are also presented here.

The densities (at 15°C) and kinematic viscosities (at 40°C) of the biodiesels are compared in Figs. 2 and 3, respectively. The densities of virgin oil biodiesels (R100, S100, C100, T100, and P100) and waste cooking oil methyl ester (W100) are close each other and approximately 5% higher than that of D100. However, the difference between D100 and SOW100 reached to 9.5 %. The higher density results in higher mass flowrate of the fuel in the engine.

The kinematic viscosities of biodiesel from virgin oils (R100, S100, C100, T100, P100) are close each other and below $5mm^2/s$. However, the viscosity of W100 is slightly higher than $5mm^2/s$. SOW100 has the highest viscosity which is approximately seven times higher than viscosity of diesel fuel No. 2. Therefore, it can not be advised to use SOW100 alone in a diesel engine.

In addition the changes of dynamic viscosities of the fuels with temperature range from 5°C to 40°C are compared in Fig. 4. It is shown in the figure that the differences between the viscosities of biodiesels and the viscosity of D100 increase as the temperature decreases. Again, SOW100 shows fairly high viscosity at low temperatures.

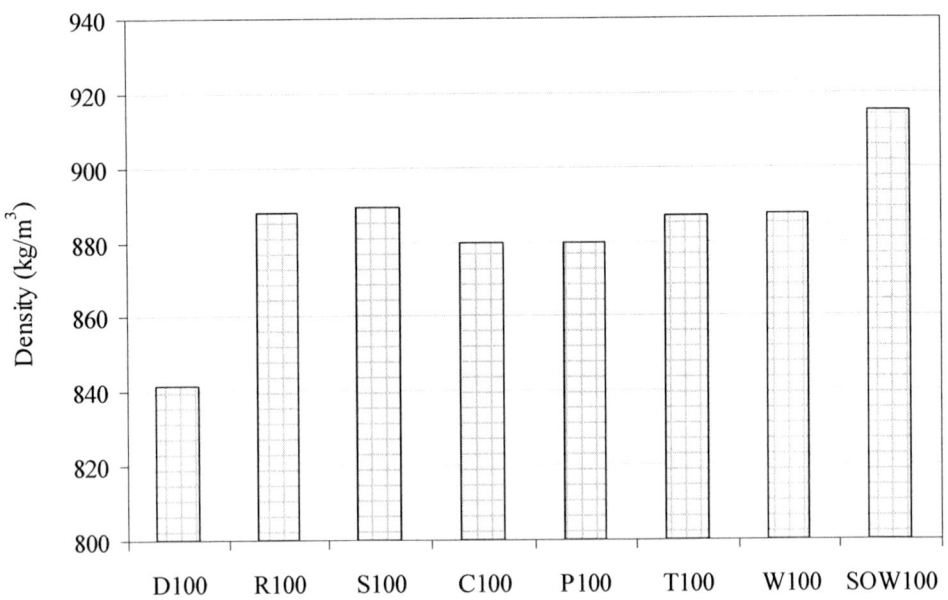

Figure 2. Density values of fuels used in the tests.

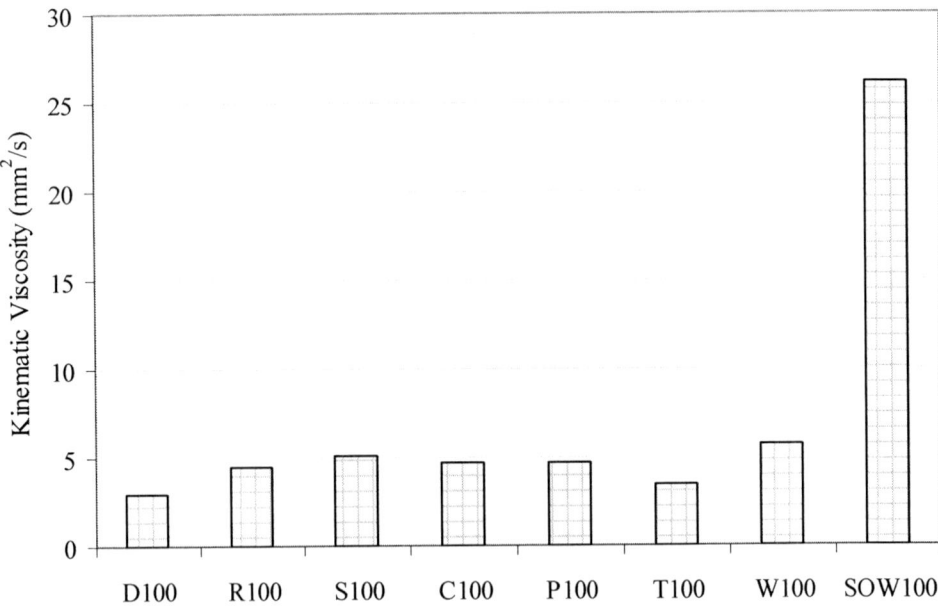

Figure 3. Kinematic viscosity values of fuels used in the tests.

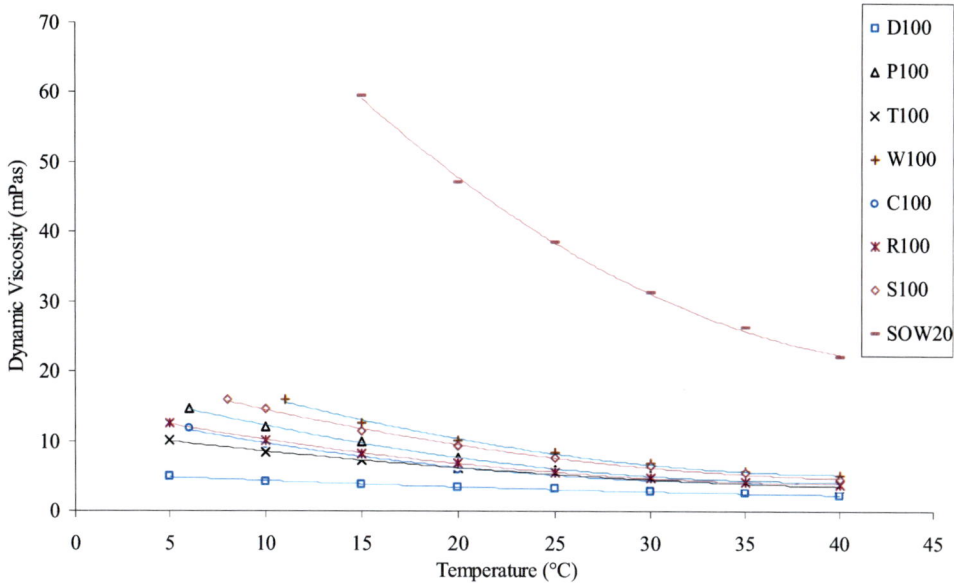

Figure 4. Change of dynamic viscosity values of the fuels with temperature.

3.2. The Engine Performance and Emission

The effects of different biodiesel additions on the performance and emission of the engine running at full load are introduced in this section. Although it is not shown here, the power variation depending on the biodiesel content in the blend was also examined using different biodiesels and it was found that the optimum power was obtained with the blends containing approximately 20 % (in volume) biodiesel. It should be pointed out that the biodiesel content in the blends is very important. After a certain amount of biodiesel content, the power reduction occurs due to the lower heating value and the higher viscosity. Although some biodiesels' viscosities are fairly close to the diesel fuel's viscosity, the reduction in power and efficiency with higher content of biodiesel in the blend is inevitable due to lower heating value (Kalam et al., 2003). If the biodiesel content is continued to increase, the power and efficiency of the blend will be lower than those of diesel fuel. The detailed information about this concept can be found in Usta et al. (2005) and Usta (2005b). This was more important for W100 and SOW100, because their viscosities are higher than that of diesel fuel and higher viscosity causes the bad atomization. Therefore, the seven different biodiesels were blended with diesel fuel No.2 in 20% (in volume).

These blends and Diesel fuel No.2 were tested in the engine. Although the measurements were recorded at 1500, 2000, 2200, 2500 and 3000 rpm, only the measurements at 3000rpm are compared here for the sake of clarity. In addition it is better to say that the maximum torque was occurred at 2200 rpm for both diesel fuel No.2 and the blends.

The effects of the seven different biodiesel additions (20% in volume) on the engine power at 3000 rpm engine speed are shown in Fig. 5. As it is seen on the figure, although the heating values of the blends are 10-11% less than that of diesel fuel, no any significant difference was determined in the engine power. The change of the power is in $\mp$1% for different blends. This can be explained with three main reasons. Firstly, since the fuel is pumped to the engine cylinders on volumetric basis, slightly higher density of the blends causes larger mass flow rate for the same fuel volume. Secondly, since the viscosities of biodiesels are higher than that of diesel fuel, more viscous blends provide less internal leakage in the fuel pump (Lang et al., 2001; Wagner et al., 1984; Al-Widyan et al., 2002). Thirdly, the biodiesels contain approximately 10-12 % (in weight) oxygen, this oxygen helps more complete combustion, thereby increasing the power (Kalam et al., 2003; Agarwal and Das, 2001).

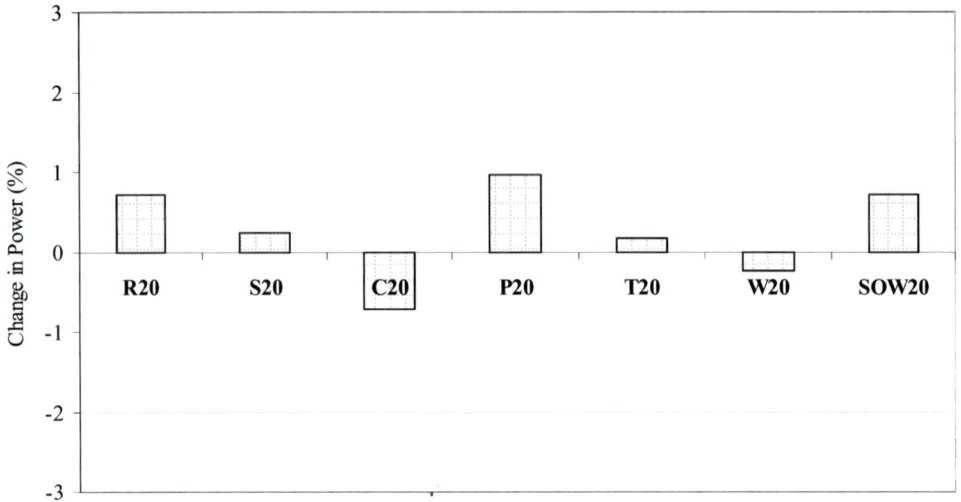

Figure 5. Change in power with respect to D100.

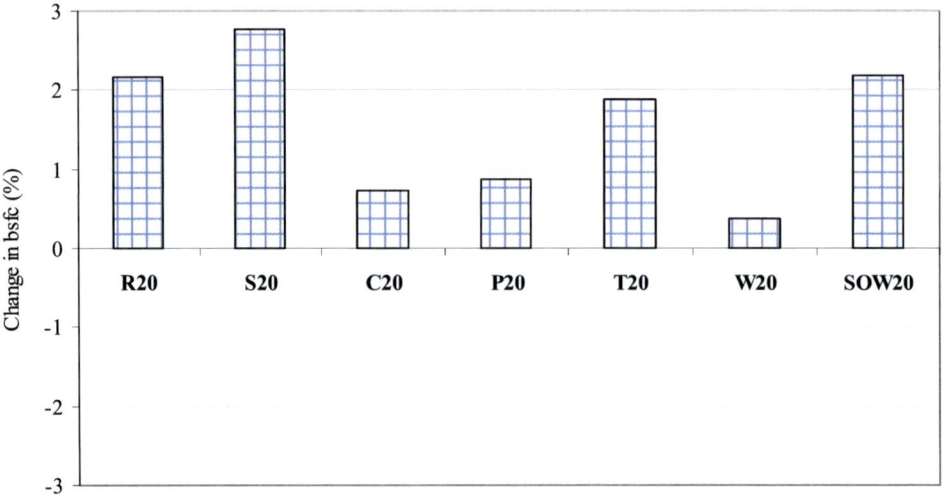

Figure 6. Change in bsfc with respect to D100.

In general, brake specific fuel consumption (bsfc) of a diesel engine depends on fuel specific gravity, viscosity, heating value and volumetric fuel injection system The fuels with lower heat content and higher density results in slightly higher bsfc with respect to the diesel fuel (Kalam et al., 2003). In the present study, bsfc values of the blends increased in the range of 0.2%-2.6% as it is shown in Fig. 6 as in (Bari et al., 2002; Kalam et al., 2003; Nwafor, 2003; de Almeida et al., 2002; Machacon et al., 2001). Highest increase in bsfc appeared with S20, meanwhile the lowest increase occurred with W20.

The thermal efficiency of a diesel engine is inversely proportional to bsfc and heating value of fuel. The bsfc values of the blends were slightly higher than diesel fuel No.2, while the heating values of the blends are lower than that of diesel fuel No.2. The thermal efficiencies obtained with the blends are compared in Fig. 7. Although the thermal efficiencies of C20, P20, T20, and W20 are slightly higher than that of diesel fuels, the efficiencies of R20, S20 and SOW20 are slightly lower than that of diesel fuel. It can be said that 20 % biodiesel addition also did not make any significant change in the thermal efficiency.

Cetane number of a biodiesel is very important specification. A biodiesel having a slightly lower cetane number results in longer ignition delay and slower burning rate (Nwafor at al. 2000) and longer ignition delay causes higher exhaust gas temperature. Late combustion in expansion stroke causes higher exhaust and lubrication oil temperatures. Also, there is another concept related to the biodiesel usage. Some biodiesels may contain some constituents which have higher boiling points are not sufficiently evaporated during the main combustion phase. They continue to burn in the late combustion phase resulting lower thermal efficiency and higher exhaust temperature (Yu et al., 2002).

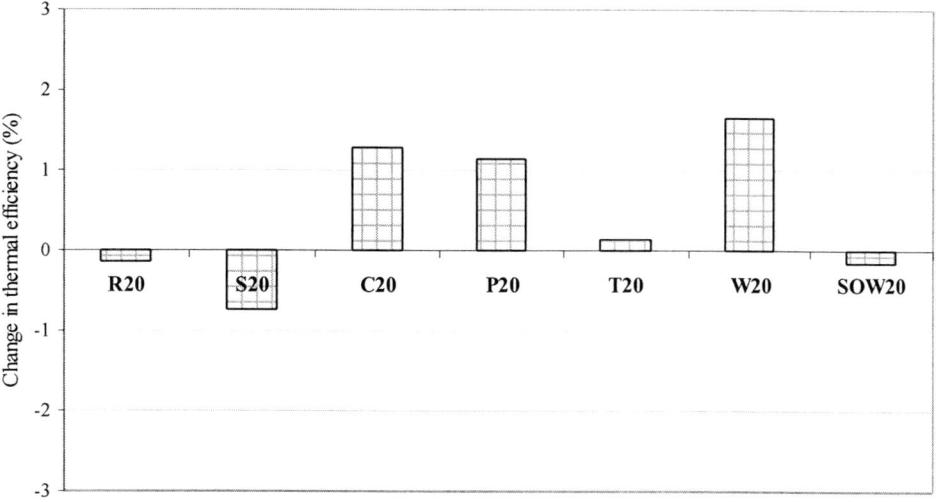

Figure 7. Change in thermal efficiency with respect to D100.

The exhaust gas temperatures of the fuels are shown in Fig. 8. It can be said that the change of the temperature is not significant. The temperature of the diesel fuel No.2 was measured as 560°C at 3000 rpm. Although R20, S20 and P20 resulted in slightly higher temperatures, the other blends caused the lower temperatures. The lowest temperature was obtained with T20 as 531°C. The results may imply that the addition of the biodiesel did not change the ignition delay.

Figure 9 shows the lubrication oil temperatures of the blends which are in general lower than that of diesel fuel in the range of 0 - 2°C. There was no significant difference in the lubrication oil temperatures with 20% biodiesel additions. This was supported with the measurements of cooling water inlet and exit temperatures of the engine. It was found that the difference between water inlet and outlet temperatures of the engine cooling system was not affected with the biodiesel addition.

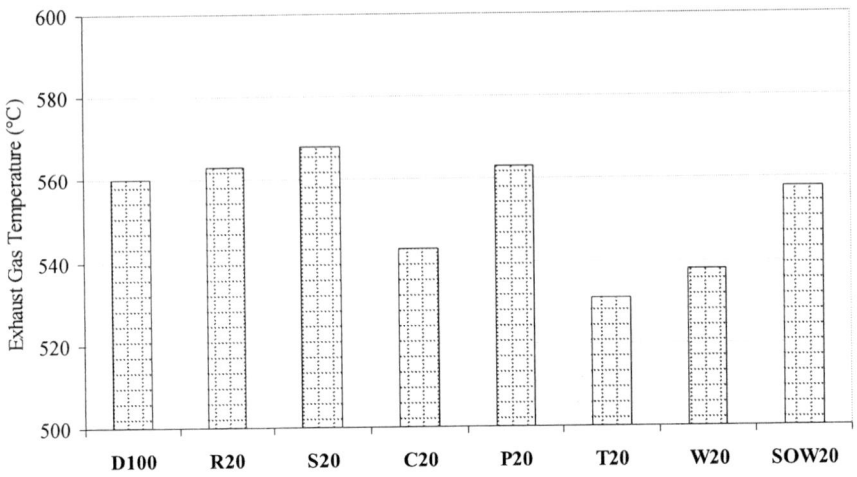

Figure 8. Comparison of the exhaust gas temperatures.

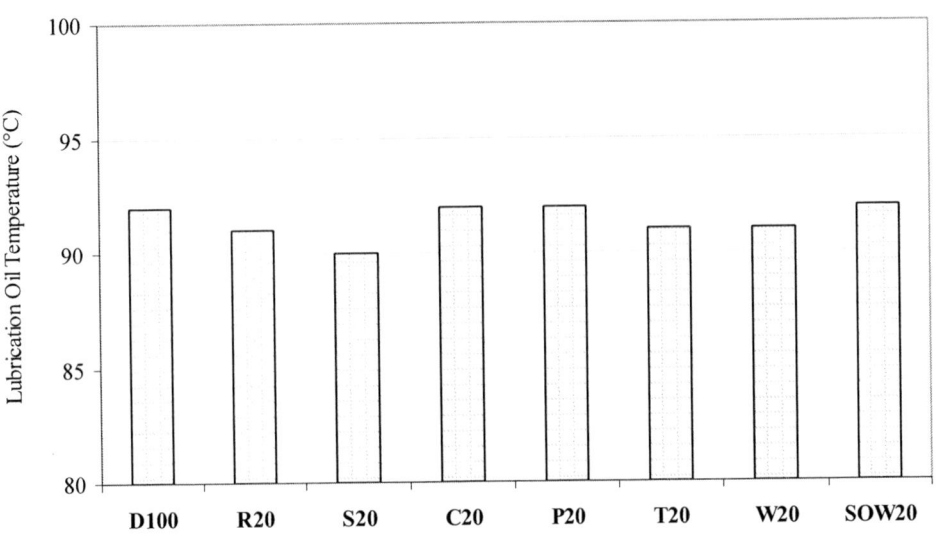

Figure 9. Comparison of the lubrication oil temperatures.

Although 20% biodiesel addition did not change the performance significantly, the emissions were affected with the biodiesel additions. One of the important emissions from CI engines is smoke. It was determined that biodiesel addition resulted in reduction in smoke emission due to complete combustion (Fig. 10) (Kalam et al.,2003; AlWidyan et al., 2002; Monyem et al., 2001). The smoke reduction reached to 32% with S20 and C20. Meanwhile the reductions in CO emission are shown in Fig. 11. The maximum reduction was determined as 35% with W20. These reductions may be explained with extra fuel oxygen. Soot and CO compete for the available oxygen in the rich combustion regime at the full engine load. In addition although the details are not shown here, at partial loads, there is no appreciable difference between the fuels due to the dominant premixed lean combustion with excess oxygen (de Almeida et al., 2002; Monyem et al., 2001).

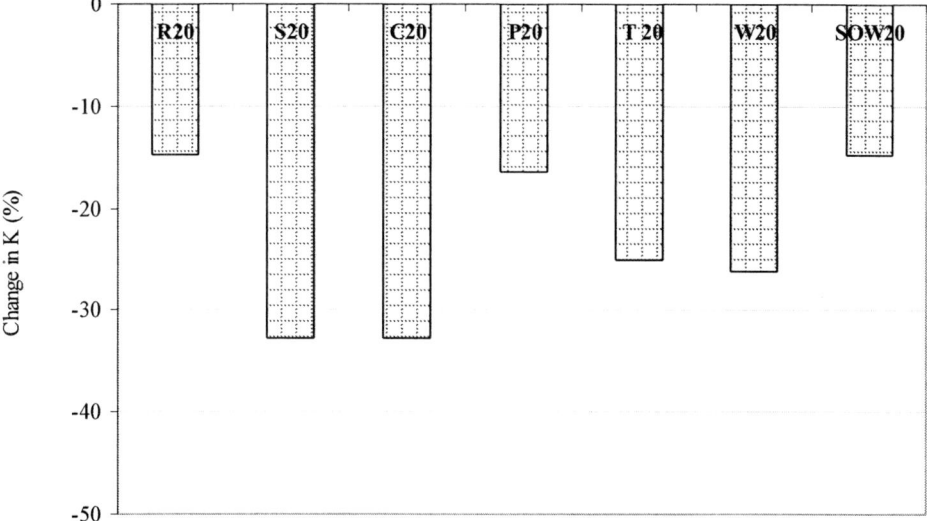

Figure 10. Change in smoke emission.

Since the biodiesels contain fairly low sulphur compared to the diesel fuel (Chang et al., 1996; Giannelos at al., 2002), it was expected to decrease in SO_2 emission. Figure 12 shows the change in SO_2 emission due to the biodiesel addition. The reduction in SO_2 emissions was higher than the expected value. Since the blend includes 20% biodiesel, SO_2 reduction was expected around 20 %. However, the measured reduction reached up to 45% at full load, similar to the results found by Dorado et al. (2003).

Although smoke, CO and SO_2 emissions decreased with the addition of the biodiesels, NO_x emission was slightly increased as it is shown in Fig. 13. It is known that three factors mainly affect NO_x emission. These are oxygen concentration, combustion temperature and time. Especially, at full load, the higher temperatures of combustion chamber and the presence of fuel oxygen causes higher NO_x emission (Gonzalez Gomez et al., 2000; Dorado et al., 2003; Yu et al., 2002). The increases in NO_x emission are in 1-6%. The maximum increase in NO_x was obtained with T20 as 5.6 %. It was also determined that the increase was fairly negligible at partial loads. Increasing of NO_x exhaust emissions is an important problem facing biodiesel usage. The biodiesels which are higher cetane numbers are correlated with reduced NO_x emissions. However, this may not always hold for all types of engine technologies (Knothe et al., 2003).

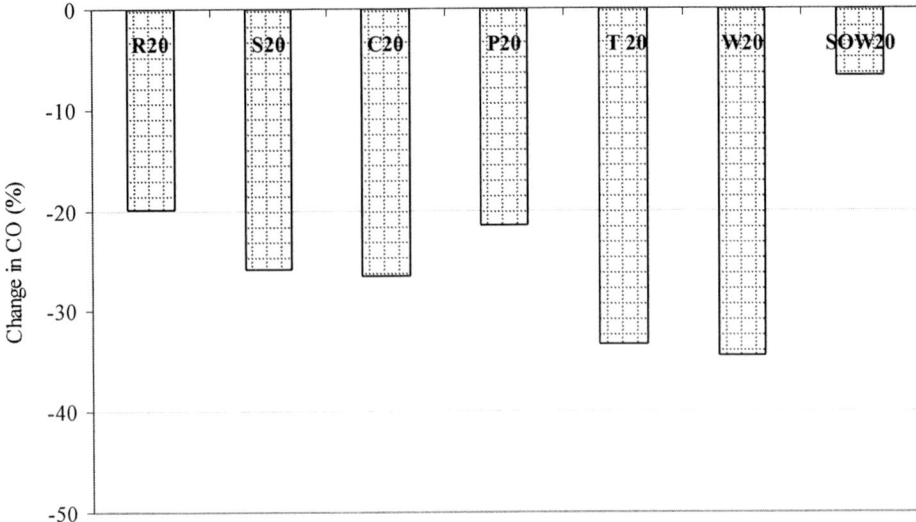

Figure 11. Change in CO emission.

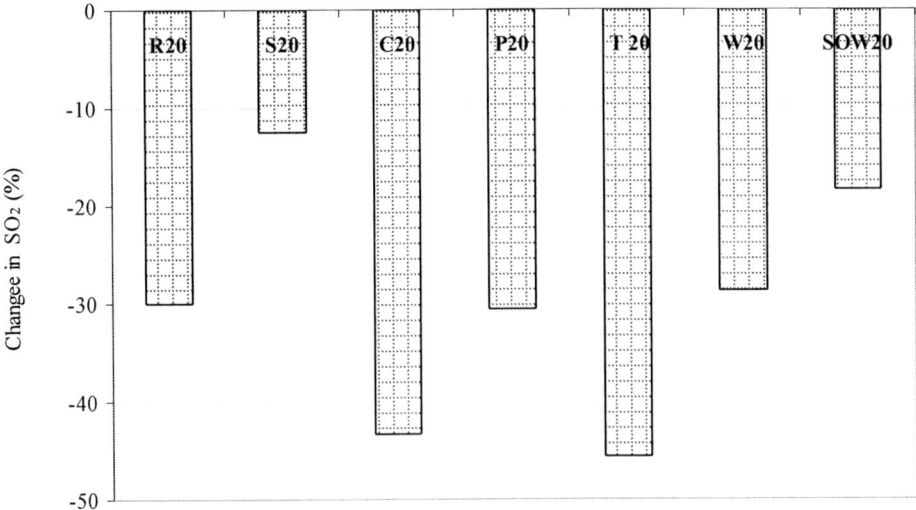

Figure 12. Change in SO_2 emission at 3000 rpm speed.

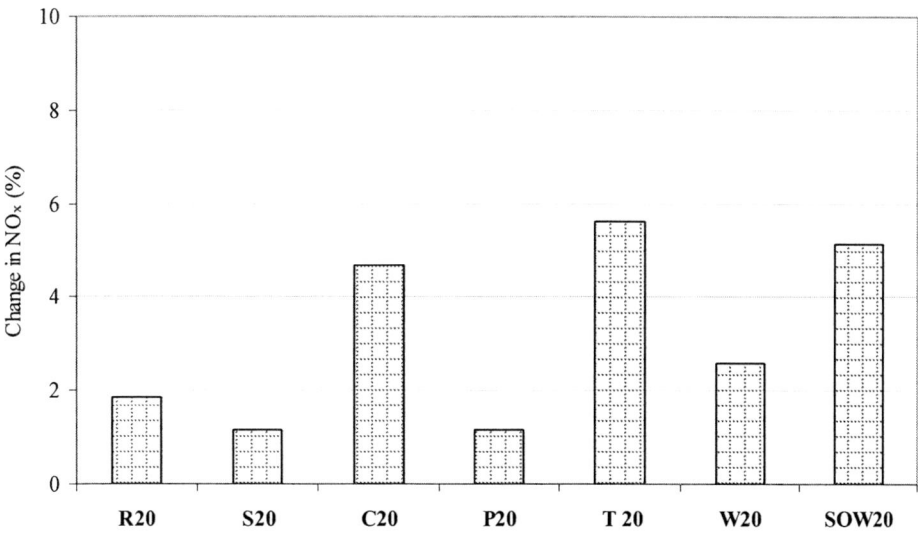

Figure 13. Change in NO_x emission at 3000 rpm speed.

4. CONCLUSIONS

In this study, seven different biodiesels were produced, and tested in a diesel engine as alternative renewable fuels. The experimental results are described as follows:

1. In addition to biodiesel production from virgin oils and waste cooking oils, it is possible to produce biodiesels from the soap stocks which have high free fatty acid content. However, the viscosity of the biodiesel produced from soapstock is fairly higher than that of diesel fuel, especially at lower temperatures.
2. The power, brake specific fuel consumption, thermal efficiency, exhaust gas temperature and lubrication oil temperature as performance parameters were examined with the addition of 20 % (in volume) seven different biodiesels. It was determined that although the heating values of biodiesels are lower than that of the diesel fuel, the biodiesel addition (20% in volume) did not cause any significant variation in the engine performance.
3. The addition of biodiesel decreased smoke and CO emissions due to the fact that biodiesels contain approximately 11- 12 % oxygen by weight, and this fuel borne oxygen helps to oxidize the combustion products in the cylinder, especially in rich region. There was a significant SO2 reduction with blends due to lower sulphur content of the biodiesel. NOx emissions slightly increased due to the presence of fuel oxygen and higher combustion temperature with the blends at full load. However, the increase is lower than 6%.
4. In these short term experiments, no obvious wear or effect on the diesel engine components was observed. However, the effects of biodiesel usage on the engine components and the lubrication oil are the subject of an ongoing project in the university.

As a result of all of the findings mentioned above, it can be concluded that biodiesels can be partially (20%) substituted for diesel fuel No.2 at most operating conditions in terms of performance parameters and emissions without any engine modification and preheating of the blends.

REFERENCES

Agarwal AK, Das LM. Biodiesel development and characterization for use as a fuel in compression ignition engines. *Journal of Engineering for Gas Turbine and Power, Transactions of ASME* 2001;123:440-447.

Alcantara R, Amores J, Canoira L, Fidalgo E, Franco MJ, Navarro A. Catalytic production of biodiesel from soybean oil, used frying oil and tallow. *Biomass and Bioenergy* 2000; 18(6):515-527.

Altin R, Çetinkaya S, Yücesu HS. The potential of using vegetable oil fuels as fuel for diesel engines. *Energy Conversion and Management* 2001;42(5):529-538.

Al-Widyan MI, Al-Shyoukh AO. Experimental evaluation of the transesterification of waste palm oil into biodiesel. *Bioresource Technology* 2002;85(3):253-256.

Al-Widyan MI, Tashtoush G, Abu-Qudais M. Utilization of ethyl ester of waste vegetable oils as fuel in diesel engines. *Fuel Processing Technology* 2002;76:91-103.

Antolín G, Tinaut FV, Briceño Y, Castaño V, Pérez C, Ramírez AI. Optimisation of biodiesel production by sunflower oil transesterification. *Bioresource Technology* 2002; 83(2):111-114.

Bari S, Lim TH, Yu CW. Effects of preheating of crude palm oil (CPO) on injection system, performance and emission of a diesel engine. *Renewable Energy* 2002;27:339–351.

Canakci M, Van Gerpen J. Biodiesel production from oils and fats with high free fatty acids. *Transaction of the ASAE* 2001; 44(6): 1429-1436.

Chang DYZ, Van Gerpen JH, Lee I, Johnson LA, Hammond EG, Marley SJ. Fuel properties and emissions of soybean oil esters as diesel fuel. *J. Am. Oil Chem. Soc.* 1996; 73:1549–55.

Chang DYZ, Van Gerpen JH. Fuel properties and engine performance for biodiesel prepared from modified feedstocks. *Society of Automotive Engineers Paper* No. 971684, SAE, Warrendale, PA, 1997.

Crabbe E, Nolasco-Hipolito C, Kobayashi G, Sonomoto KIA. Biodiesel production from crude palm oil and evaluation of butanol extraction and fuel properties. *Process Biochemistry* 2001;37(1):65-71.

Curran HJ, Fisher EM, Galude PA, Marinov NM, Pitz WJ, Westbrook C.K, Layton DW, Flynn PF, Durrett RP, Zur Loye AO, Akinyemi OC, Dryer FL. Detailed chemical kinetic modeling of diesel combustion with oxygenated fuels. *Society of Automotive Engineers Paper* No. 2001-01-0653971684, SAE, Warrendale, PA, 2001.

Darnoko D, Cheryan M. Kinetics of palm oil transesterification in a batch reactor. *JAOCS* 2000;77(12):1263-1267.

de Almeida SCA, Belchiora CR, Nascimentob MVG, Vieirab LDSR, Fleuryb G. Performance of a diesel generator fuelled with palm oil. *Fuel* 2002;81:2097-2102.

Demirbaş A. Biodiesel from vegetable oils via transesterification in supercritical methanol. *Energy Conversion and Management* 2002;43:2349–2356.

Demirbaş A. Biodiesel fuels from vegetable oils via catalytic and non-catalytic supercritical alcohol transesterifications and other methods: A survey. *Energy Conversion and Management* 2003;44(13):2093-2109.

Dorado MP, Ballesteros E, Arnal JM, Gómez J, López FJ. Exhaust emissions from a diesel engine fueled with transesterified waste olive oil. *Fuel* 2003;82(11):1311-1315.

Ergeneman M, Ozaktas T, Karaosmanoglu F, Arslan HE. Ignition delay characteristics of some Turkish vegetable oil-diesel fuel blends. *Petroleum Science and Technology* 1997a; 15(7-8):667-683.

Ergeneman M, Ozaktas T, Cigizoglu KB, Karaosmanoglu F, Arslan E. Effect of some Turkish vegetable oil-diesel fuel blends on exhaust emissions. *Energy Sources* 1997b;19(8):879-885.

Giannelos PN, Zannikos F, Stournas S, Lois E, Anastopoulos G. Tobacco seed oil as an alternative diesel fuel: Physical and chemical properties. *Industrial Crops and Products* 2002;16:1–9.

Gonzalez Gomez ME, Howard-Hildige R, Leahy JJ, O'reilly T, Supple B, Malone M. Emission and performance characteristics of a 2 Litre Toyota diesel van operating on esterified waste cooking oil and mineral diesel fuel. *Environmental Monitoring and Assessment* 2000;65:13–20.

Graboski MS, McCormick RL, Alleman TL, Herring AM. The Effect of Biodiesel Composition on Engine Emissions from a DDC Series 60 Diesel Engine, Final Report February 2003, NREL/SR-510-31461, National Renewable Energy Laboratory, Golden, Colorado, USA.

Graboski MS, McCormick RL. Combustion of fat and vegetable oil derived fuels in diesel engines. *Prog. Energy Combust. Sci.* 1998;24:125-164.

Haas MJ, Bloomer S, Scott K. Simple, high-efficiency synthesis of fatty acid methyl esters from soapstock. *JAOCS* 2000; 77(4): 373-379.

Haas MJ, Scott KM, Alleman TL, McCormick RL. Engine performance of biodiesel fuel prepared from soybean soapstock: A high quality renewable fuel produced from a waste feedstock. *Energy and Fuels* 2001; 15(5): 1207-1212.

Haas MJ, Foglia TA. Cheaper feedstocks for biodiesel. Industrial Bioprocessing 2002; 24(5): 4-5.

Haas MJ, Michalski PJ, Runyon S, Nunez A, Scott KM. Production of FAME from acid oil, a by-product of vegetable oil refining. *JAOCS* 2003; 80(1): 97-102.

Kalam MA, Husnawan M, Masjuki HH. Exhaust emission and combustion evaluation of coconut oil-powered indirect injection diesel engine. *Renewable Energy* 2003;28:2405-2415.

Kalligeros S, Zannikos F, Stournas S, Lois E, Anastopoulos G, Teas CH and Sakellaropoulos F. An investigation of using biodiesel/marine diesel blends on the performance of a stationary diesel engine. *Biomass and Bioenergy* 2003;24(2):141-149.

Karaosmanoglu F. Vegetable oil fuels: A review. *Energy Sources* 1999;21:221-231.

Karaosmanoğlu F, Kurt G, Özaktas T. Long term CI engine test of sunflower oil. *Renewable Energy* 2000;19:219-221.

Karaosmanoğlu F., Tuter M., Gollu E., Yanmaz S., Altıntıg E., Fuel Properties of Cotton seed Oil. *Energy Sources* 1999; 21:821- 828,

Kitamura T, Ito T, Senda J, Fujimoto H. Extraction of the suppression effects of oxygenated fuels on soot formation using a detailed chemical kinetic model. *JSAE Review* 2001a;22:139-145.

Kitamura T, Ito T, Senda J, Fujimoto H. Detailed chemical kinetic modeling of diesel spray combustion with oxygenated fuels. Society of Automotive Engineers Paper No. 2001-01-1262, SAE, Warrendale, PA, 2001b.

Knothe G., Matheaus A.C., Ryan T.W. Cetane numbers of branched and straight-chain fatty esters determined in an ignition quality tester. *Fuel* 2003;82: 971–975.

Kumar MS, Ramesh A, Nagalingam B. An experimental comparison of methods to use methanol and Jatropha oil in a compression ignition engine. *Biomass and Bioenergy* 2003;25:309–318.

Lang X, Dalai AK, Bakhsi NN, Reaney MJ, Hertz PB. Preparation and characterization of bio-diesels from various bio-oils. *Bioresource Technology* 2001;80:53-62.

Ma F, Hanna MF. Biodiesel production: A review. *Bioresource Technology* 1999;70:1-15.

Machacon HTC, Shiga S, Karasawa T, Nakamura H. Performance and emission characteristics of a diesel engine fueled with coconut oil-diesel fuel blend. *Biomass and Bioenergy* 2001; 20: 63-69.

Marshall W, Schumacher LG, Howell S. Engine exhaust emissions evaluation of a cummins 110e when fueled with a biodiesel blend. Society of Automotive Engineers Paper No. 952363, SAE, Warrendale, PA, 1995.

Monyem A, Van Gerpen JH. The effect of biodiesel oxidation on engine performance and emissions. *Biomass and Bioenergy* 2001;20:317-325.

Monyem A, Van Gerpen JH, Canakci M. The effect of timing and oxidation on emissions from biodiesel-fueled engines. *Transactions of the ASAE* 2001;44(1):35-42.

Nwafor OMI. The effect of elevated fuel inlet temperature on performance of diesel engine running on neat vegetable oil at constant speed conditions. *Renewable Energy* 2003;28: 171-181.

Nwafor OMI, Rice G, Ogbonna AI. Effect of advanced injection timing on the performance of rapeseed oil in diesel engines. *Renewable Energy* 2000;21:433-444.

Ozaktas T. Compression ignition engine fuel properties of a used sunflower oil-diesel fuel blend. *Energy SourA□s* 2000;22(4):377-382.

Ozaktas T, Cigizoglu KB; Karaosmanoglu F. Alternative diesel fuel study on four different types of vegetable oils of Turkish origin. *Energy Sources* 1997;19(2):173-181.

Piazza G.J. and Foglia T.A., Rapeseed oil for oleochemical usage *Eur. J. Lipid Sci. Technol.* 2001; 103: 450–454.

Pramanik K. Properties and use of jatropha curcas oil and diesel fuel blends in compression ignition engine. *Renewable Energy* 2003;28(2):239-248.

Srivastava A, Prasad R. Triglycerides-based diesel fuels. *Renewable and Sustainable Energy Reviews* 2000;4:111-133.

Tickell J, Tickell K. From the fryer to the fuel tank: The complete guide to using vegetable oil as an alternative fuel. Green Teach Pub., Sarasota, FL, 2nd ed. 1999, 66-67.

Tomasevic AV, Siler-Marinkovic SS. Methanolysis of used frying oil. *Fuel Processing Technology* 2003;81(1):1-6.

Usta N., Öztürk E., Can Ö., Conkur E.S., Nas S., Çon A.H., Can A.Ç., Topcu M. Combustion of biodiesel fuel produced from hazelnut soapstock/waste sunflower oil mixture in a diesel engine. *Energy Conversion and Management 2005*; 46: 741-755.

Usta N. Use of tobacco seed oil methyl ester in a turbocharged indirect injection diesel engine, *Biomass and Bioenergy* 2005a; 28: 77-86.

Usta N. An experimental study on performance and exhaust emissions of a diesel engine fuelled with tobacco seed oil methyl ester, *Energy Conversion and Management* 2005b; 46: 2373–2386.

U.S. Department of Agriculture, National Agricultural Statistics Service. Statistics of Cotton, Tobacco, Sugar Crops and Honey, Chapter II.
http://www.usda.gov/nass/pubs/agr03/03_ch2.pdf, 2004.

Vicente G, Coteron A, Martinez M, Aracil J. Application of the factorial design of experiments and response surface methodology to optimize biodiesel production. *Industrial Crops and Products* 1998;8(1):29-35.

Wagner L, Clark S, Schrock M. Effects of soybean oil esters on the performance, lubricating oil, and wear of diesel engines. Society of Automotive Engineers Paper No. 841385, SAE, Warrendale, PA, 1984.

Williamson AM, Badr O. Assessing the viability of using rape methyl ester (RME) as an alternative to mineral diesel fuel for powering road vehicles in the UK. *Applied Energy* 1998; 59(2-3):187-214.

Yu CW, Bari S, Ameen A. A comparison of combustion characteristics of waste cooking oil with diesel as fuel in a direct injection diesel engine. *Proc. Instn. Mech. Engrs. Part D: J. Automobile Engineering* 2002;216:237-243.

In: Progress in Biomass and Bioenergy Research
Editor: Steven F. Warnmer, pp. 177-204

ISBN: 1-60021-328-6
© 2007 Nova Science Publishers, Inc.

Chapter 7

NEW MATERIALS FROM LIGNIN

Carlo Bonini and Maurizio D'Auria

Dipartimento di Chimica, Universita' della Basilicata,
Via N. Sauro 85, 85100 Potenza, Italy

ABSTRACT

Lignin, obtained through steam explosion from straw, was completely characterized via elemental analysis, gel permeation chromatography, ultraviolet and infrared spectroscopy, ^{13}C and ^{1}H nuclear magnetic resonance spectrometry.

Lignin powder was used for the preparation of blends with low-density polyethylene (LDPE), linear low-density polyethylene (LLDPE), high-density polyethylene (HDPE) and atactic polystyrene (PS).

The obtained blends are processable through the conventional techniques used for thermoplastics; the modulus slightly increases for most lignin-polymer blends, while the tensile stress and elongation reduce. Moreover, lignin acts as a stabilizer against the UV radiation for PS, LDPE and LLDPE.

Polyurethanes were obtained treating steam exploded lignin from straw with 4,4'-methylenebis(phenylisocyanate), 4,4'-methylenebis(phenylisocyanate) – ethandiol, and poly(1,4-butandiol)tolylene-2,4-diisocyanate terminated. The obtained materials were characterized by using gel permeation chromatography, infrared spectroscopy and scanning electron microscopy. Differential scanning calorimetry analysis showed a T_g at -6 °C, assigned to the glass transition of the poly(1,4-butandiol) chains. The presence of ethylene glycol reduced the yields of the polyurethanes. The use of the prepolymer gave the best results in polyurethanes formation. Steam exploded lignin was used as starting material in the synthesis of polyesters. Lignin was treated with dodecanoyl dichloride. The products were characterized by using gel permeation chromatography, infrared spectroscopy, ^{13}C and ^{1}H nuclear magnetic resonance spectrometry, and scanning electron microscopy.

INTRODUCTION

Lignin is a natural amorphous polymer, which, together with cellulose and hemicellulose, is one of the main constituents of wood. It is generally obtained, as a by-product in the paper production, through the separation from the cellulose fibers. Its structure depends of the kind of process used for delignification [1]. Due to its phenolic nature, many chemical modifications have been studied. For example, it has been used as a main chain on which other synthetic polymer chains can be grafted [2]. Moreover, due to the presence of the phenolic groups it is expected that it can increase the oxidation, thermal and light stability of polymeric materials. It is a low-density, low abrasive and low-cost material, features that could be interesting in its use as filler instead of inorganic fillers [3]. With certain polymers, in suitable formulations, it can give partially or completely biodegradable composites [4]. All these features are attractive from the industrial point of view, nevertheless rarely lignin has been used to obtain new materials; in fact only in the last years some trials have been made to use it as a thermoplastic. In the literature some papers have been published concerning the use of lignin as a stabilizer for plastics and rubbers, where it acts as antioxidant or modifier of the mechanical properties [4,5]. The main papers concern blends of lignin with polyvinylacetate [6], where it improves some mechanical properties, with polyethylene and polypropylene, where it acts as a stabilizer against degradation reactions [3,4,7-9], with polyvinylchloride [10], in which it increases the yielding stress, with biodegradable polyesters and polyamides, where, in the first case it improves the impact resistance, and in the second the rigidity [11]. In these works wood lignin is always used. In fact, a wide quantity of lignin is obtained as a byproduct of the paper industry and traditionally has been used as a fuel to produce energy. On the other hand, non-used lignin constitutes a major environmental problem, therefore it would be important to find for it new applications.

Several uses of lignin in the synthesis of new materials have been reported. In particular, lignin has been used as raw material in the preparation of polyurethanes [12-20] and in the synthesis of graft copolymers [21-32]. Furthermore, several examples of the synthesis of polyesters from lignin have been reported [33-37]. Polyesters can be used in the formulation of polyurethane coatings.

The separation of lignin from cellulose is made by means of the pulping process in the paper industry, which presents environmental problems and give altered lignin.. Nevertheless, another important source of lignin exists, straw, a very diffuse and very low cost agricultural residue and nowadays new delignification processes exist that have been largely developed mainly due to the environmental problems related to the pulping process and that allow to obtain a less altered lignin, in particular the Steam Explosion Process.

Steam explosion is a technology useful for the treatment of every lignocellulosic material. In the steam explosion saturated vapour at high pressure is used to rapidly warm the biomass in a digester. The biomass is maintained at the desired temperature (130 – 180 °C) for a short time: during this period the hemicellulose is hydrolysed and dissolved. At the end of this period, an explosive decompression gives rise to a loss of water from the cells (due to the immediate evaporation of water) and the cleavage of cellular structures.

In our work we use mainly straw lignin, largely present in our region, due to the large abundance of agricultural crops, separated through the Steam Explosion process, developed at the ENEA of Trisaia (Matera, Italy) to obtain blends with synthetic polymers (polyethylenes and polystyrene), which have been chosen for their large commercial diffusion, low cost and processing temperature range close to that of lignin. The aim of our work is the obtainment of low cost materials, of potential interest in the field of packaging, health care products, agricultural films, disposable objects, to find new applications for lignin, a natural abundant and low cost polymer, for which, nowadays, only a small market exists.

CHARACTERIZATION OF LIGNIN

To perform our experiments we used a steam-exploded lignin from straw. The results of the elemental analysis were C: 62.13, H: 5.88; N: 1.26; S: 0.00; O: 30.73%. We analyzed the presence of carbon and hydrogen in order to characterize the lignin, but also the presence of both nitrogen, as a marker of the presence of proteins in the lignin, and sulphur, as a marker of the presence of sulphate lignin. The presence of sulphur in our sample was not detected. The elemental analysis allowed us to give the molecular weight of the lignin expressed in phenylpropanoid (C_9) units. In our case, the molecular formula was $C_9H_{10.22}O_{3.34}$ with a molecular weight of 172. Elemental analysis showed that this sample was highly oxidised with a large amount of oxygen in the molecular formula.

The distribution of the molecular weights of acetylated lignin was obtained by using GPC: it gave M_n = 3509, M_w = 15096, and M_z = 40966. These data confirmed the evidence that the steam explosion process induced a strong destructuration in the lignin structure giving samples with relative low molecular weight.

The UV spectrum of the lignin from straw was recorded in DMF. It showed absorption at 231 nm (D = 16.8 L g^{-1} cm^{-1}). We recorded the differential spectrum carrying out the spectrumin1 M NaOH vs.the standard solution in DMF. These data allowed us to give the amount in mEq g^{-1} of some structural features in the lignin sample. We could give the amount of syringyl and guaiacyl phenols (Type I), the amount of phenols containing conjugated double bonds (i.e. HO-Ar-CH=CH-CH$_2$OH, Type II), and the amount of stilbenic phenols (Type IV). We found in our sample Type I (0.43mEq g^{-1}) and Type IV (0.12 mEq g^{-1}).

The infrared spectrum showed absorptions at 1702 (carbonyl stretching), 1655 (C=O stretching in aryl ketones), 1605 and 1513 (aromatic stretching), 1459 (C-H bending in methyl and methylenic groups), 1424 (aromatic vibration coupled with C-H bending in plane), 1330 (C-H bending in plane in syringyl and guaiacyl rings substituted on C-5), 1220 (C-C, C-O, and C=O stretching), 1123 (C-H bending in syringyl units and C-O stretching in secondary alcohols), 1030 (C-H bending in plane in guaiacyl units and C-O stretching in primary alcohols), and 840 cm^{-1} (aromatic C-H bending out of plane).

The ^{13}C NMR spectrum of lignin from straw gave signals at δ 173 (C=O), 153 (C-3/C-3'in 5-5'etherified units), 148 (C-4 in etherified guaiacyl units), 145 (C-4 in β-O-4 non etherified guaiacyl units), 138 (C-1 in β-O-4 etherified syringyl units), 135 (C-4 in β-O-4 etherified and non etherified syringyl units), 133 (C-1 in β-O-4 non etherified guaiacyl units), 130 (C-2/C-6 in benzoate), 120 (C-6 in etherified and non etherified guaiacyl units), 115 (C-5 in etherified and non etherified guaiacyl units), 112 (C-2 in guaiacyl units), 111 (C-2 in guaiacyl-guaiacyl stilbenes), 105 (C-2/C-6 in syringyl units), 87 (C-β in β-O-4 *threo* syringyl units), 72 (C-α in β-O-4 *erythro* guaiacyl and syringyl units), 60 (C-γ in β-O-4 *erythro* and *threo* syringyl and guaiacyl units), 56 (methoxy groups), and 34-20 ppm (CH$_3$ and CH$_2$ in saturated chains). The ^{13}C NMR spectrum was compatible with the presence of both guaiacyl and syringyl units. Furthermore, the ^{1}H NMR spectrum on acetylated lignin showed signals at δ 1.9-2.0 (aliphatic acetates), 2.18 and 2.30 (aromatic acetates), 2.6 (benzyl protons in 3-aryl-1-propanol units), 3.8 (methoxy groups), 6.6 (aromatic protons in syringyl units), 6.9 (aromatic protons in guaiacyl units), and 7.6 ppm (aromatic protons ortho to carbonyl groups). ^{1}H NMR spectrum showed the presence of signals due to the presence of aromatic acetates, in agreement with the differential UV spectrum showing the presence of large amount of phenolic hydroxy groups in the structure (0.55 mEq g^{-1}). The Figure 1 represents the ESEM image of the lignin from straw.

Figure 1. ESEM of lignin from straw.

PREPARATION OF BLENDS

The blends have been prepared through the extrusion of the synthetic polymers and the lignin, in a nitrogen atmosphere, in a single screw and single chamber Brabender extruder with the following parameters: a. extruder diameter: 30 mm, b. L/D ratio: 25, c. compression ratio 1/3.5. In the choice of the temperatures (in the range from 160 °C to 190 °C), the degradation of lignin, observed trhough TGA, see below, has to be taken into account.

In Table 1 all the samples with the legends used to identify them are indicated. The blend composition is expressed as weight % of lignin

The DSC curve of lignin from 60 °C to 200 °C at 20 °C/min is reported in Figure 2.

The scan shows a large endotherm near 100 °C, indicating the presence of water, then a gradual increase of the specific heat at about 161 °C, that can be easily attributed to the glass transition of lignin.

Table 1. Samples examined in the preparation of blends

SAMPLE	LEGEND
Lignin	Lignin
Flexirene (LLDPE)	Flexirene
Blend containing 10% by weight of lignin and 90% by weight of Flexirene	LF10
Blend containing 20% by weight of lignin and 80% by weight of Flexirene	LF20
Riblene (LDPE)	Riblene
Blend containing 10% by weight of lignin and 90% by weight of Riblene	LR10
Blend containing 20% by weight of lignin and 80% by weight of Riblene	LR20
Lupolen (HDPE)	Lupolen
Blend containing 10% by weight of lignin and 90% by weight of Lupolen	LLU10
Blend containing 20% by weight of lignin and 80% by weight of Lupolen	LLU20
Polystyrene Aldrich	PSA
Blend containing 10% by weight of lignin and 90% by weight of Polystyrene Aldrich	LPSA10
Blend containing 20% by weight of lignin and 80% by weight of Polystyrene Aldrich	LPSA20
Polystyrene Dow	PSD
Blend containing 20% by weight of lignin and 80% by weight of Polystyrene Dow	LPSD20
Blend containing 30% by weight of lignin and 80% by weight of Polystyrene Dow	LPSD30

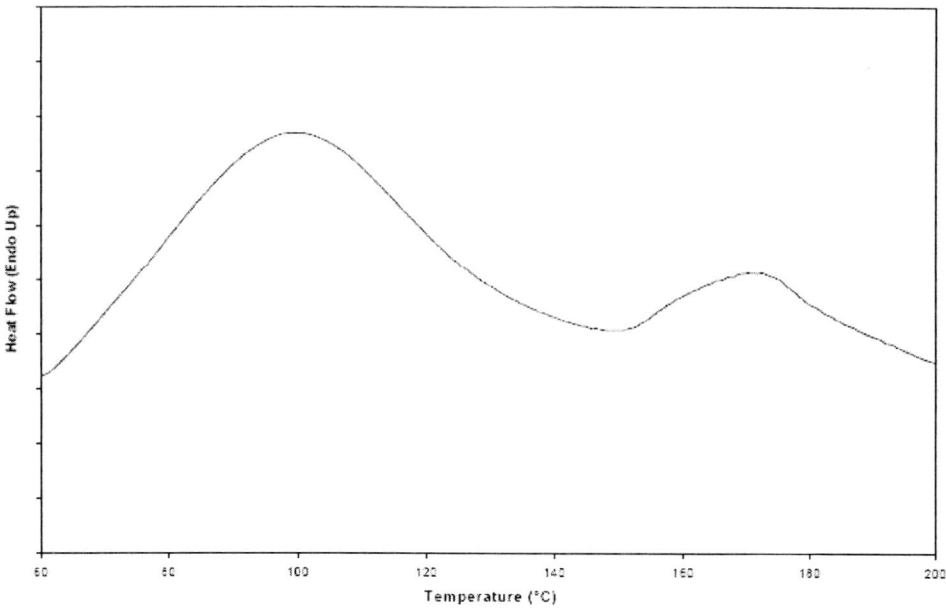

Figure 2. DSC scan of lignin from 60 °C to 200° C at 20 °C/min.

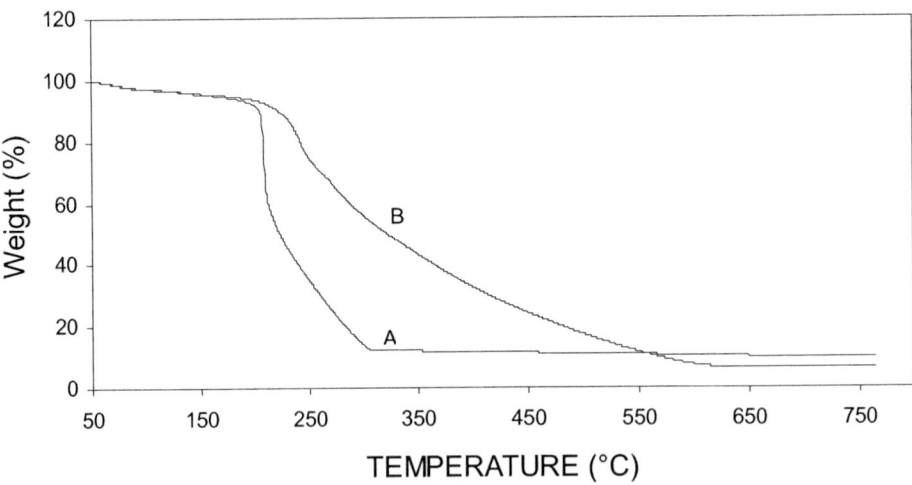

Figure 3. TGA scan of lignin in air (A) and in nitrogen (B) at 20 °C/min.

The subsequent heating of lignin, after cooling from 200 °C to 50 °C at the same scanning rate, does not show any change in the specific heat. Such behavior is similar to what reported in the literature for various wood lignins, that, after thermal treatment, do not show any glass transition [6]. This can be interpreted by admitting that the thermal treatment determines an increase in the glass transition temperature (T_g) related to a decrease in the chain flexibility, likely due, in turn to a chemical or physical (*e.g.* hydrogen bond formation) crosslinking process.

The TGA scans in air and in a nitrogen atmosphere, reported in Figure 3, reveal a weight loss of about 5% at 100 °C, due to the adsorbed water, followed by the complete degradation of lignin at about 210 °C in air and 230 °C in nitrogen.

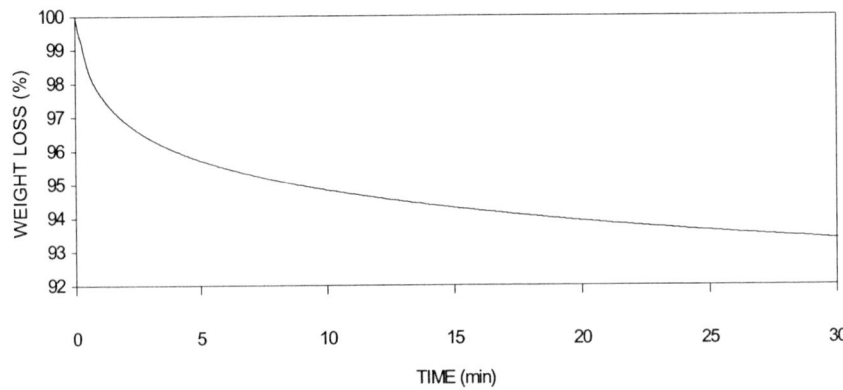

Figure 4. TGA isotherm of lignin at 190 °C for 30 min.

In order to careful evaluate the degradation of lignin, which has to be taken into account to define the processing conditions, various isotherms in a nitrogen atmosphere have been performed. As an example Figure 4 shows the isotherm obtained after taking the sample at 190°C, keeping it for 30 min at this temperature.

A weight loss of 6.62% takes place.

The TGA measurements allowed to establish 200 °C as the maximum temperature to be used for the extrusion, as at higher temperatures a too large degradation of lignin occurs (the complete degradation takes place at 230 °C in a nitrogen atmosphere), taking into account the weight loss of 6.62% already at 190 °C after 30 min and of the further heat developing in the extruder because of the friction.

The Karl-Fisher titration revealed a water content of 4.3% in weight, confirming the result obtained through TGA.

The Melt Flow Index (MFI) of the neat polymers and of the blends has been evaluated, in order to have a qualitative indication of the effect of the addition of lignin on the viscosity, and hence on the processing properties, of the materials. Figure 5 shows the trend of MFI as a function of the lignin content in the blend.

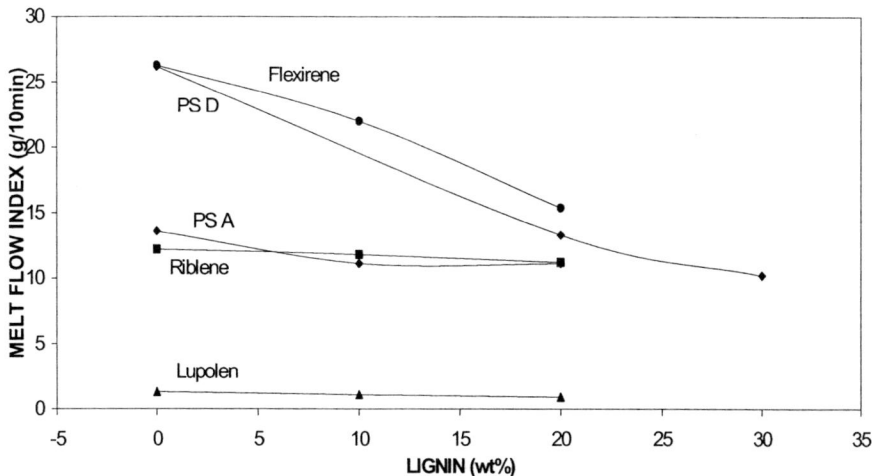

Figure 5. Dependence of MFI as function of the lignin content in the blends with Flexirene, Riblene, Lupolen, PSA and PSD.

In all cases a decrease of the MFI is obtained and, as a consequence, an increase of the viscosity, as expected, on increasing the lignin content in the blends.

Tensile tests have been performed to study the effect of lignin on the basic mechanical properties of the polyethylenes and polystyrenes.

In Table 2 for all the samples are reported the values of the maximum tensile stress and of elongation at breaking evaluated in the tensile tests.

The maximum tensile stress decreases for all samples on increasing the lignin content. In particular for Flexirene a strong decrease occurs (about 68%) passing from the neat polymer to the blend LF10, then a slight further decrease takes place when the content of lignin is increased to 20%.

Passing from neat Riblene to the blend LR10 a very strong reduction of the maximum tensile stress (about 75%) is observed, value that remains constant after a further addition of lignin in LR20.

Table 2. Maximum tensile stress (σ_{max}) and elongation (ε_{max}) at breaking

Sample	σ_{max} (MPa)	ε_{max} (%)
Flexirene	23.0	2000
LF10	6.1	112.5
LF20	4.4	104.2
Riblene	20.8	1391.8
LR10	5.2	54.2
LR20	5.5	58.3
Lupolen	34.5	1962.5
LLU10	18.0	38.5
LLU20	18.8	41.8
PSA	29.6	6.6
LPSA10	12.0	11.8
LPSA20	4.6	-
PSD	28.8	5.6
LPSD20	8.5	2.5
LPSD30	5.9	-

For Lupolen the decrease of the maximum tensile stress is about half than in the previous cases for the blend containing 10% of lignin (about 32%). Nevertheless, on increasing the content of lignin up to 20% a further decrease of 16% occurs.

For PSA and PSD a decrease of the maximum tensile stress is observed upon increasing the lignin content.

Also the elongation at breaking decrease upon the addition of lignin. This decrement is particularly high for the sample containing polyethylene, while is lower for those containing polystyrene.

The stress-strain curves for some samples are reported in Figure 6. In Figure 6A the curve of Flexirene is reported as an exemplum of polyethylene behavior. Similar curves are recorded for Riblene ans Lupolen, which therefore are not reported. The usual curve observed in the tensile mode at a constant deformation rate of plastic material is obtained. In the case of lignin-modified samples (FL10 is reported as an exemplum in Figure 6B), the curves appear strongly modified. The fracture thakes place at much lower stress values and the elongation at breaking is much lower. The shape of the curve is different with respect to that of neat polyethylene. It shows a ductile-stable trend, with a first linearly increasing trend, then a deviation from linearity occurs before the strain reaches its maximum value. Then, there is a region of plastin deformation with constancy of stress on elongation, then the failure takes place up to the breaking of the specimen.

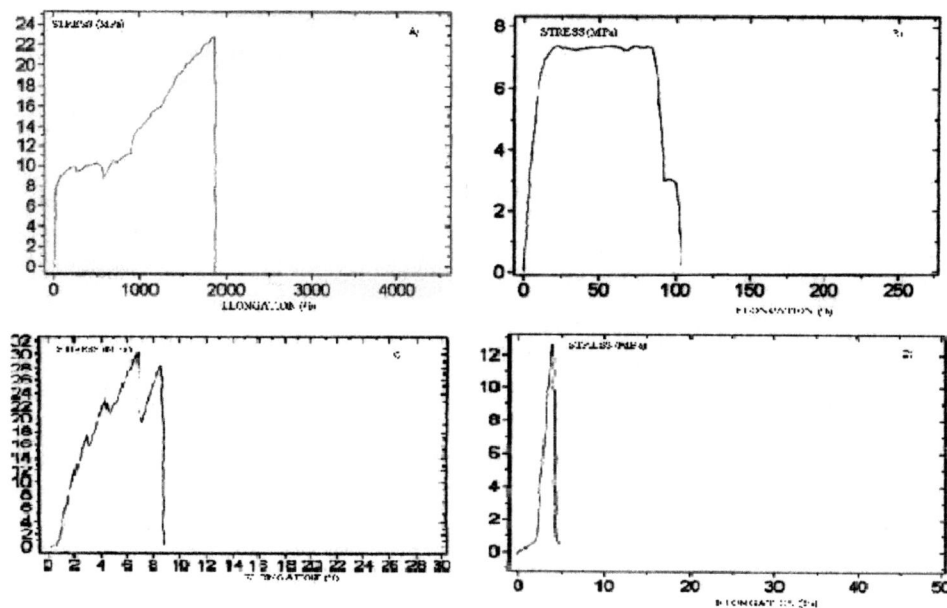

Figure 6. Stress-strain curves for neat Flexirene (A), LF10 (B), neat PSA (C) and LPSA 10 (D).

In Figure 6C the stress-strain curve for PSA is reported as an example of poly(styrene)-containing samples. A fragile fracture, typical of glassy polymers is observed. In particular, the fracture is fragile-unstable, in that it propagates in an unstable manner through the specimen until the elastic energy storerd in the sample itself begins not sufficient for the further growth and the fracture stops.

This phenomenon repeats itself more times and it is possible to evidence on the curve the points of the crake initiation and stop up to the fracture of the specimen. For the modified sample LPSA10 (Figure 6D) the fracture is fragile too, but occurs at much lower value of the stress.

The thermal propertied of the samples have been evaluated by DSC. In Table 3 the glass transition temperatures (T_g), the melting temperatures (T_m) and the heat of fusion, obtained by DSC for the neat polymers and for the blends are reported.

In Figure 7 the DSC scan of Lupolen (a) and for the blends LLU10 (b) and LLU20 (c) are reported, as exempla of poly(ethylene) containing samples. From Table 3 and Figure 7 it can be observed that the presence of lignin scarcely influences the thermal behavior of neat Lupolen. In fact, the melting peak, centered at 133 °C for the neat polymer (a), slightly decreases to 129 °C for both blends (b and c). The normalized heat of fusion decreases on increasing the lignin content in the blends is observed. This beahavior may point to a lower nucleation density, then to the formation of poorer and less crystals.

For the other PE-containing samples a very similar behavior, with even smaller effects, is observed.

Table 3. Melting temperatures (T$_m$) and heats of fusion (ΔH$_m$) normalized with respect to the synthetic polymer content in the blends, glass transition temperatures (T$_g$), evaluated by DSC

Sample	T$_m$(°C)	ΔH$_m$(J/g)	T$_g$ (°C)
Flexirene	111.0/119.8	63.7	
LF10	113.7/120.0	62.6	
LF20	113.7/119.9	42	
Riblene	111.5	65.5	
LR10	112.5	66.5	
LR20	112.5	56.3	
Lupolen	132.7	148.5	
LLU10	129.5	117.4	
LLU20	129.4	93.6	
PSA			99.9
LPSA10			100.6
LPSA20			97.9
PSD			95.6
LPSD20			95.9
LPSD30			94.6
Lignin			161

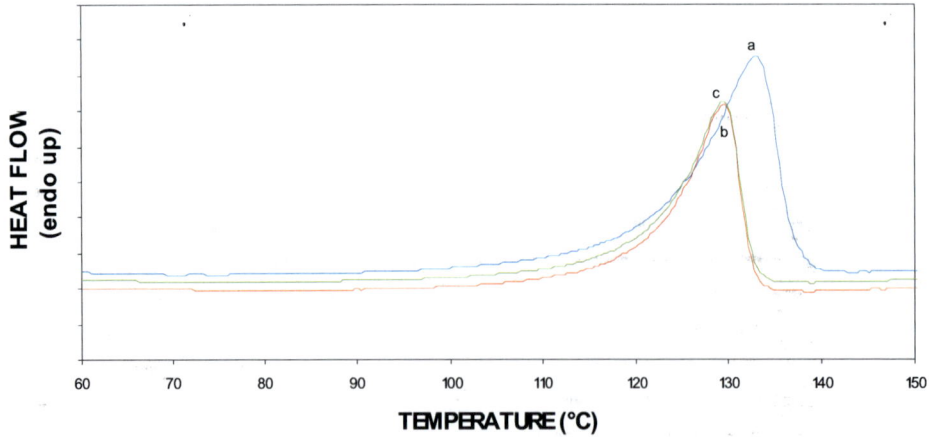

Figure 7. DSC scan from 60 °C to 150 °C at 20 °C/min for Lupolen (a), LLU10 (b), LLU20 (c).

For PSD, in Figure 8 and Table 3, we observe the glass transition at a T$_g$ around 95 °C, scarcely influenced in the blends by the presence of lignin. A similar behavior is obtained with PSA.

The dynamic mechanical properties, namely the storage component of the modulus and the loss factor have been evaluated for the neat polymers and for the blends. In table 4 the glass transition temperatures of all samples are reported together with the α-transition ones, in the case of PE-containing samples, evaluated by DMTA.

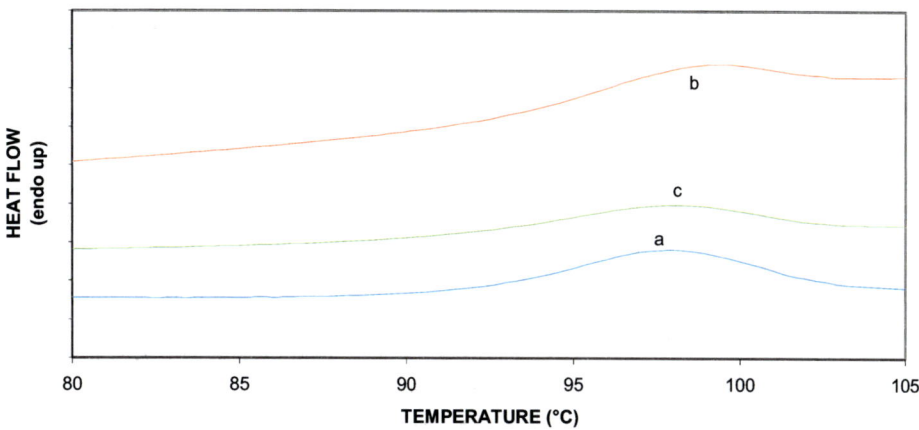

Figure 8. DSC scan from 50°C to 200 °C at 20 °C/min for PSD (a), LPSD20 (b), LPSD30 (c).

In all cases in tanδ curve two peaks are detected, the first one at lower temperature, due to the glass transition of polyethylene, the other at temperatures lower than the melting point, which corresponds to the α-transition of polyethylene and has been attributes to a crystalline relaxation due to both intralammellar and intralamellar-c schear [38].

Table 4. α-Transition temperatures (T_α) and glass transition temperatures (T_g) evaluated through DMTA·

Sample	$T_\alpha(°C)$	$T_g(°C)$
Flexirene	50.7	-113.2
LF10	71.6	-111
LF20	52	-111
Riblene	56.8	-121.6
LR10	56.8	-118
LR20	51.6	-118
Lupolen	103.8	-110.5
LLU10	97.8	-110.5
LLU20	95.8	-110.5
PSA		115
LPSA10		117.5
LPSA20		117.5
PSD		105.4
LPSD20		105.5
LPSD30		107.1

Figure 9 shows the trend of the storage component of the modulus (log E') and of the loss factor (tanδ) as a function of the temperature, for Lupolen (a), LLU10 (b) and LLU20 (c), as exempla of the PE-containing samples.

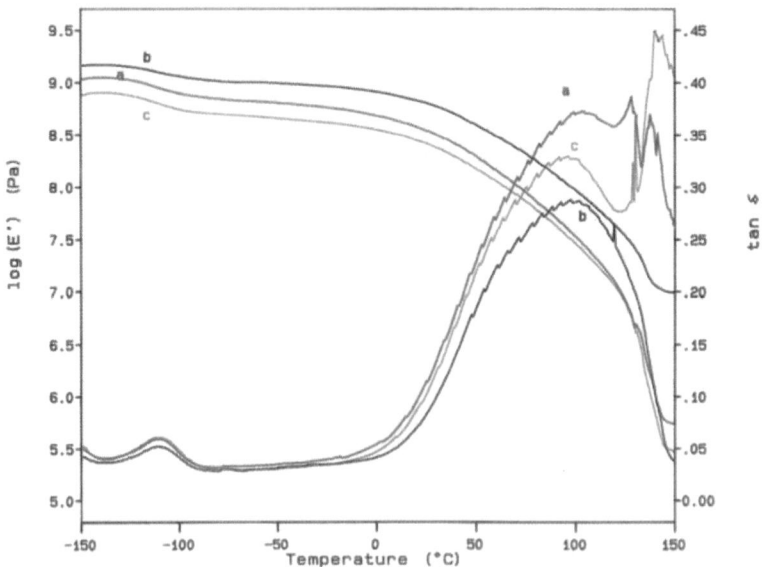

Figure 9. Storage component of the modulus (logE') and loss factor (tanδ) as a function of temperature for Lupolen (a), LLU10 (b), LLU20 (c) evaluated by DMTA.

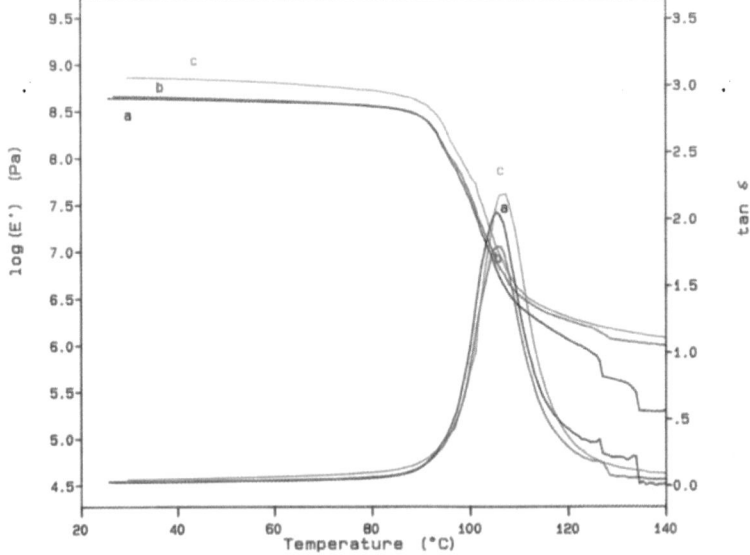

Figure 10. Storage component of the modulus (logE') and loss factor (tanδ) as a function of temperature for PSD (a), LPSD20 (b), LPSD30 (c) evaluated by DMTA.

In general, on passing from the neat polymer to the blends, a slight increase in T_g and a decrease in T_α are observed (except for Lupolen for which T_g is nearly constant at about − 110.5 °C).

In Figure 10 the trend of the storage component of the modulus (log E') and of the loss factor (tanδ) as a function of temperature are reported for PSD (a), LPSD20 (b) and LPSD30 (c), as exampla of the polystyrene-containing samples.

tan δ curve shows for such samples the presence of a single peak, attributable to the polystyrene glass transition. As far as the modulus is concerned, it generally increases, after the addition of lignin.

Morphological investigations have been carried out, by scanning electronic microscopy, examining the surface of the extruded samples, fractured in liquid nitrogen. In Figures 11-18 the ESEM micrographs of the neat polymers and of the blends containing 10% by weight of lignin and 20% in the case of PSD are reported.

The ESEM micrograph of neat flexirene (Figure 11) shows a rough, irregular surface, with evidente yielding signs, as typical for a polyethylene sample.

In the presence of lignin , sample LF10 (Figure 12), the surface appears smoother, showing a more fragile fracture. Lignin particles are clearly distributed nonuniformly within the matrix and with a large distribution of their dimensions. Such particles are characterized by an internal structure and protrude out of the matrix, revealing a poor adhesion.

The fracture surface of Riblene (Figure 13) shows a fibrillar appearance, where yielding signs can be noted.

The sample LR10 presents a rough surface (Figure 14), nevertheless the lignin particles, inhomogeneously dispersed within the matrix are completely separated, evidencing the complete absence of interfacial adhesion.

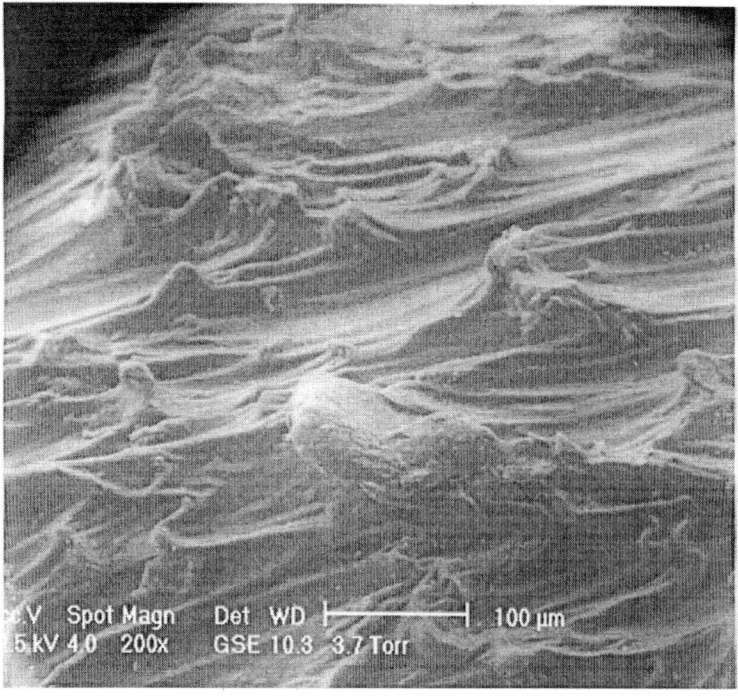

Figure 11. ESEM micrograph of the fracture surface of Flexirene.

In the ESEM micrograph of neat Lupolen (Figure 15) the surface, although quite smooth, shows clear signs of yielding.

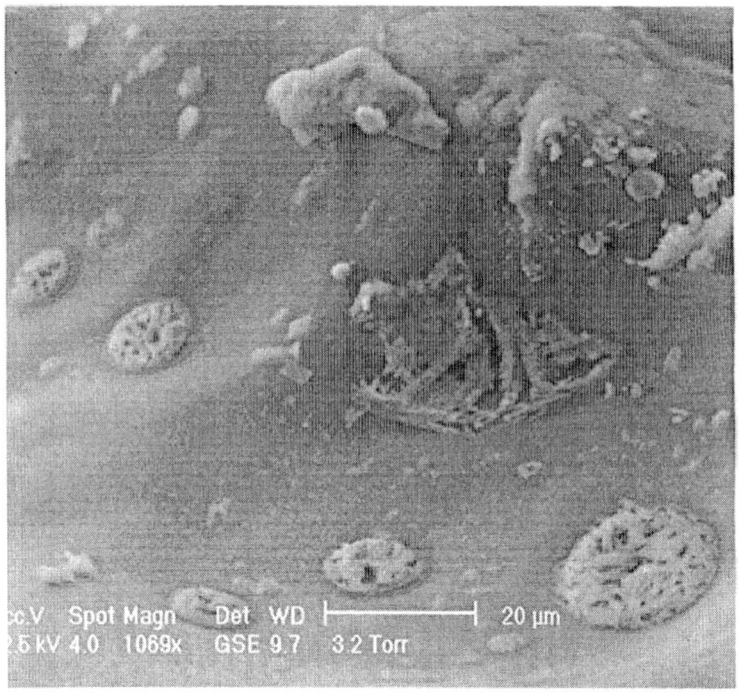

Figure 12. ESEM micrograph of the fracture surface of LF10

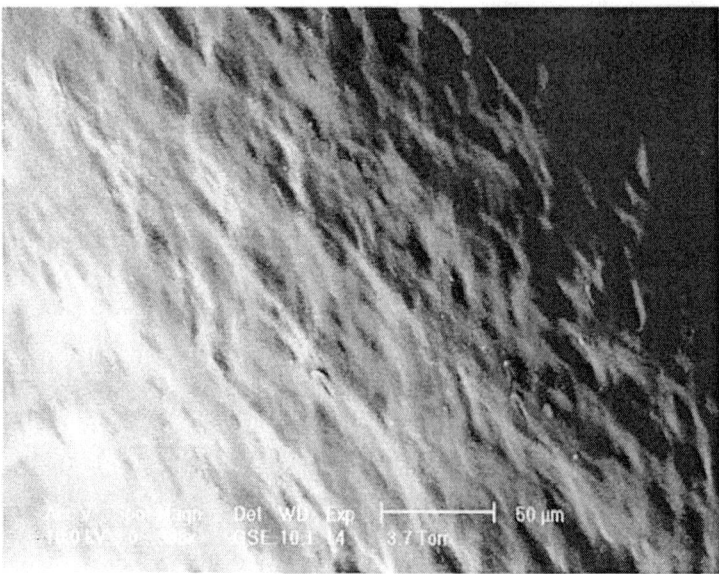

Figure 13. ESEM micrograph of fracture surface of Riblene.

The surface of the lignin-modified polymer appears rough (Figure 16). Also here some plastic deformation is present, nevertheless the lignin particles, smaller than in the other cases, are, separated from the matrix.

In the case of PSD the fracture surface, shown in Figure 17 is mirror-like, revealing an essentially fragile fracture, typical of glassy polymers.

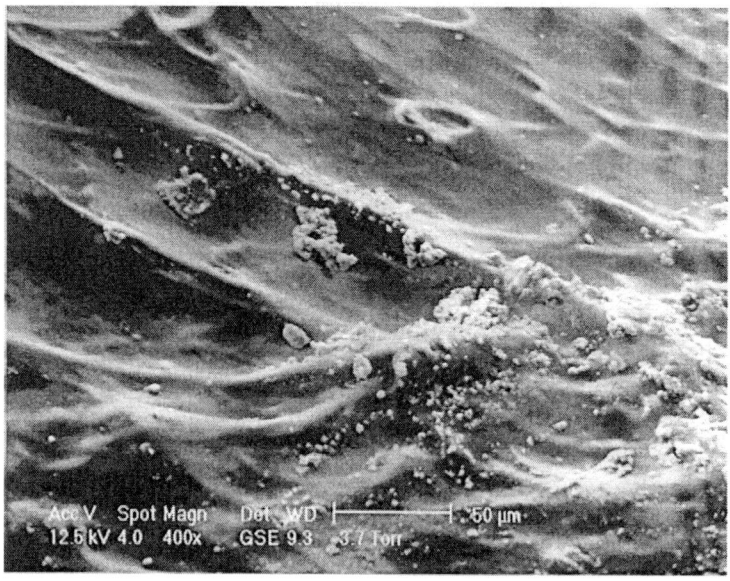

Figure 14. ESEM micrograph of the fracture surface of LR10

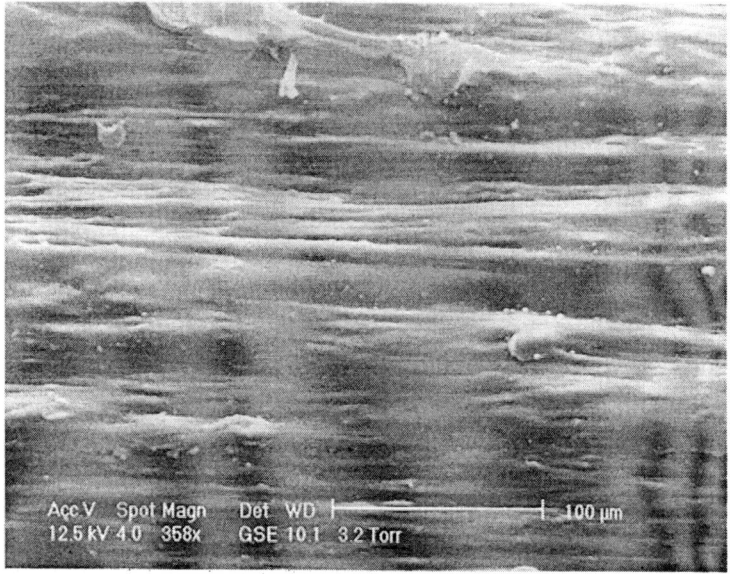

Figure 15. ESEM micrograph of the fracture of Lupolen.

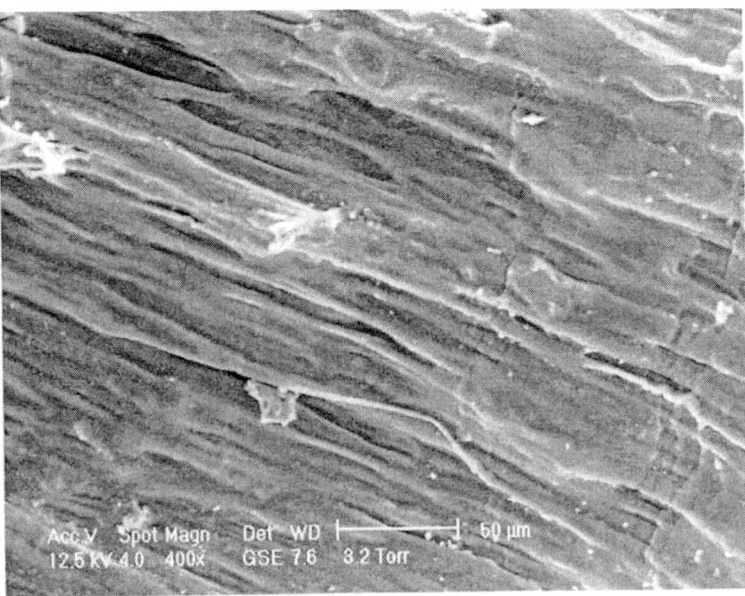

Figure 16. ESEM micrograph of the fracture surface of LLU10.

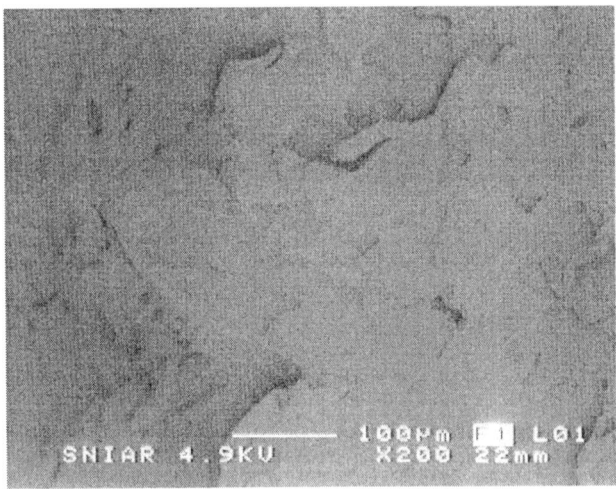

Figure 17. ESEM micrograph of the fracture surface of PSD.

In the case of the blend containing 20% by weight of lignin, holes within the matrix are observed (Figure 18), likely due to the ablation of the lignin particles, confirming the very poor adhesion between lignin and matrix.

The sample have been tested for the UV stability determination of the effect of lignin presence and content. The results of the photodegradation are reported in Figures 19-22.

As far as the neat polymers are concerned, we observe that under irradiation PSD and Flexirene undergo some degradation, Riblene remains unvaried, while for Lupolen an increase of the molecular weight occurs, that may be due to crosslinking reactions among the chains induced by the UV radiation.

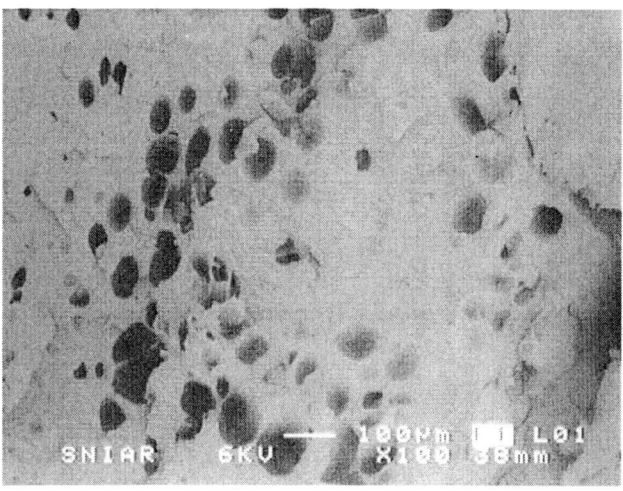

Figure 18. ESEM micrograph of the fracture surface of LPSD20.

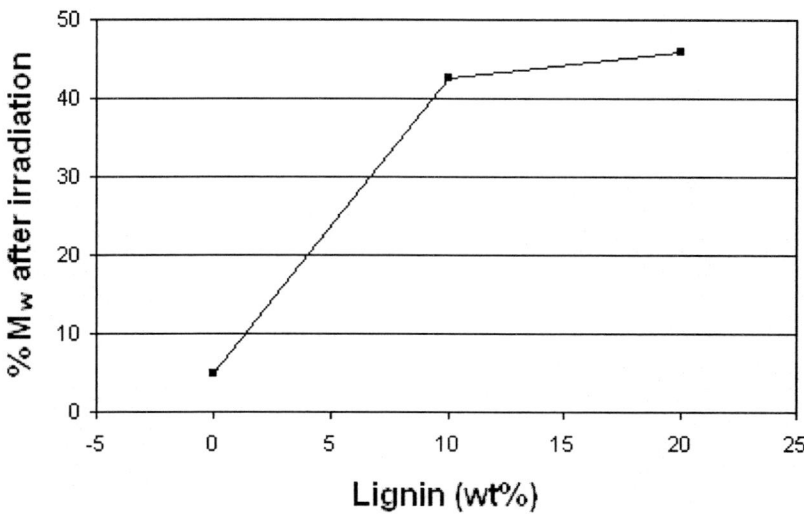

Figure 19. Percent of the weight average molecular weight hold after irradiation of blends Lignin-Flexirene with a 15W UV lamp for 48 hours.

The addition of lignin to PSD, Flexirene and Riblene causes an increase in the average molecular weight after irradiation, on increasing the content of lignin, suggesting a protecting action against photoossidation exerted by the filler. In contrast, the addition of lignin causes a neat degradation of Lupolen which clearly decomposes.

The addition of lignin to polyethylene and polystyrene does not hinder their processability. In fact, although MFI increases upon addition of lignin, the samples are always well processable. During processing, the molecular wight of polymers and their distribution can be modified, in particular the molecules undergo mechanical stress that can result in the rupture of the macromolecules themselves, with the consequent decrease of the molecular weight. In contrast, in the present case, a decrease in MFI is obtaines. It is worthy noting that such a result is in contrast with that obtained in ref. 4 for blends of polyethylene and polypropylene with wood lignin. In fact, in this case, always an increase in MFI was observed upon addition of lignin to the thermoplastic polymers. Therefore, our results appear encouraging in suggesting the use of straw-lignin as a processing stabilizer. Nevertheless, in order to understand if lignin can effectively act as a processing stabilizer for the above polymers, instead of the conventional inorganic stabilizers, which are more expensive, toxic and abrasive, the trend of MFI as a function of the number of subsequent extrusions should be performed.

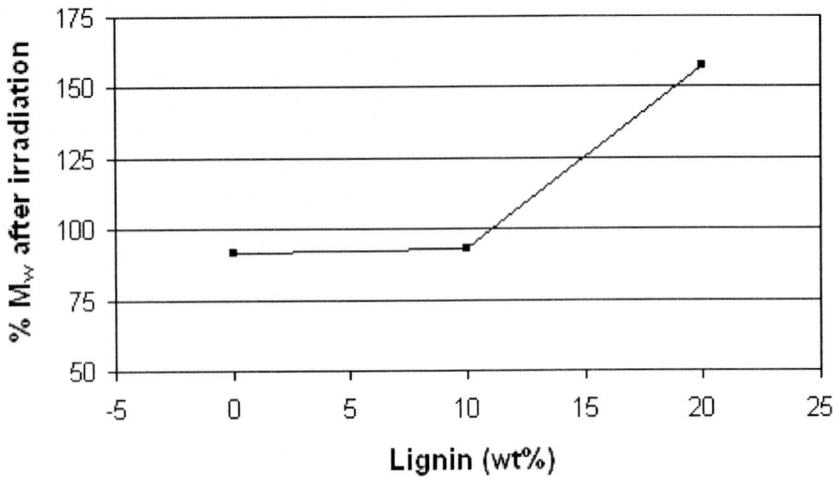

Figure 20. Percent of the weight average molecular weight hold after irradiation of blends Lignin-Riblene with a 15W UV lamp for 48 hours.

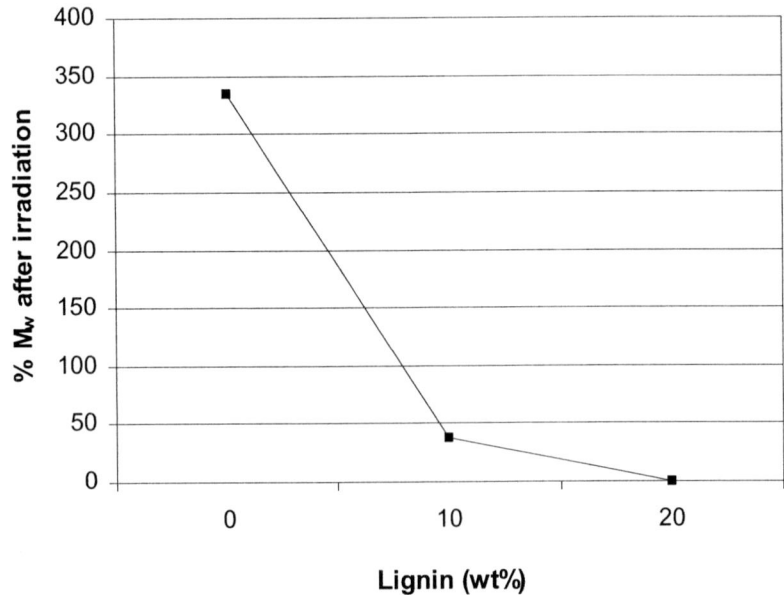

Figure 21. Percent of the weight average molecular weight hold after irradiation of blends Lignin-Lupolen with a 15W UV lamp for 48 hours.

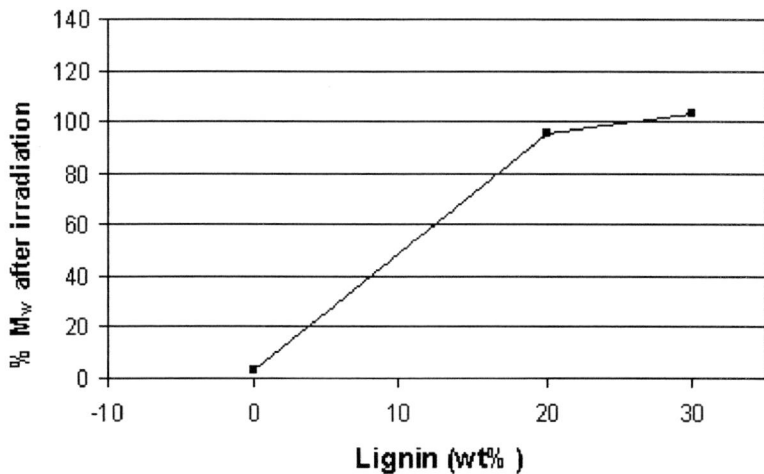

Figure 22. Percent of the weight average molecular weight hold after irradiation of blends Lignin-PSD with a 15W UV lamp for 48 hours.

As far as the mechanical properties are concerned, in all cases a decrement of the maximum tensile stress and of the elongation at breaking occurs in the blends with respect to the neat polymers. Similar results are reported in literature for non-polar thermoplastics and lignin blends [3,4] and, more generally, for polymer composites, in which a poor compatibility takes place between the two components.

Therefore, in order to evaluate the compatibility between lignin and synthetic polymers used in the blends, we have performed a series of investigations. Firstly, DSC investigations have been performed on the neat polymers and on the blends. For both the kind of blends, those containing semicrystalline polymers (polyethylenes) and those containing glassy one (polystyrenes) the DSC results point to the absence of miscibility with lignin. In fact, in the first case melting of polyethylene is scarcely affected by the addition of lignin, except for a very small lowering of the melting temperatures. For the samples containing polystyrene, the constancy of the polystyrene glass transition temperature in the blend also indicates the absence of miscibility with lignin.

The same indications are given by the DMTA analysis. In fact, from the loss factor curve as a function of temperature it can be observed that the typical transition of the neat polymer (glass and α transition for polyethylenes, glass transition for polystyrenes) appear unchanged in the blends, as far as their temperature and shape are concerned.

Therefore, the decrement of the stress and the elongation at breaking of the considered blends have to be ascribed to the poor miscibility between polar lignin and non polar synthetic polymers. In these systems rigidity is ordinarilt improved, but streght, elongation and toughness sacrified.

In fact, also in our case, as far as the modulus is concerned, it generally increases, after the addition of lignin. This result would suggest that lignin may be used as a reinforcing agent, i.e. as a filler suitable to increase such mechanical features of the material, in general rigidity, but even the dimensional stability and the shrinkage. In general, inorganic fillers are used to this target, but the lignin low abrasivity, low cost and the absence of toxicity could render it competitive with respect to these ones.

In order to further investigate the interactions between lignin and the thermoplastic polymers, we have performed morphological investigations, by ESEM, on the neat polymers and on the blends. In all cases the lignin particles are distributed in a inhomogeneous way within the polymer matrix, moreover they appear separated from the matrix itself. Often they have been ablated leaving holes within the matrix.

Therefore, the mechanical behavior can been ascribed to the poor adhesion between the nonpolar polymer and the more polar lignin used as a filler, that gives a poor stress transfer between the matrix and the filler and yielding of the lignin-polymer composites at lower stress values than for the unblended polymers. The reduction of the tensile stress can also be ascribed to a poor dispersion of the filler into the matrix. In all cases, therefore, the decrease of the mechanical properties is clearly attributable to the poor adhesion and dispersion of the lignin particles in the apolar polymeric matrix, which create defects that act as stress concentrators and make the specimen fragile.

Nevertheless, in our opinion, the observed mechanical properties constitute a lower limit, since the distribution of the lignin particles in the matrix could be improved by using a more efficient mixing technology and the adhesion by using a compatibilizing agent.

The photodegradation behavior of PSD and Flexirene clearly shows that the neat polymers are strongly degraded by the UV radiation, while Riblene remains more stable. In the lignin-containing blends we observe a much lower degradation than in the unblended polymer, while for Riblene and PSD not only a constancy of the molecular weight, which would indicate an antioxidant action by lignin, but even an increase. This could indicate the occurrence of coupling reactions between the radicals present on the polymer chains and on lignin, giving formation of branched chains. Lupolen shows a completely different behavior, with an increase of the molecular weight of the neat polymer after irradiation, likely due to branching and crosslinking. The addition of lignin induces, in contrast, high degradation. The UV stabilization performed by lignin on low-density polyethylenes (Flexirene and Riblene) and polystyrene is due to the presence of phenolic groups, which make lignin to act as a radical scavenger, inhibiting or slowering the radicalic processes of degradation. The different behavior observed in Lupolen blends may be explained taking into account that lignin (as well as other fillers, in general) distributes inhomogeneously in the polymer, in particular it concentrates in the material amorphous regions. This could explain the strong UV stabilization induced on polystyrene (amorphous), on Riblene (LDPE) and Flexirene (LLDPE), which show low crystallinity degrees, evaluated by DSC, (0.22 for both). The negative effect detected in Lupolen could lie in the much higher crystallinity (0.51) of such polymer (HDPE). Therefore lignin is poorly mixed with it, in contrast it can act as initiator in its degradation, as reported in the literature for polypropylene [4].

SYNTHESIS OF POLYURETHANES

Lignin and 4,4'-methylenebis(phenylisocyanate) were suspended in THF for 5 h in the presence of catalytic amount of stannous octoate. After solvent evaporation, the product was maintained in an oven at 72 °C overnight (Table 5).

Table 5. Polyurethanes from steam-exploded lignin

Entry	Lignin [g]	Diisocyanate[a] [g]		mEq	Ethandiol [g]	Yield [g]
		A	B			
1	1.000	0.138		2.04		0.826
2	1.000	0.200		3.06		0.806
3	1.000	0.270		4.08		0.830
4	1.000	0.138		2.04	0.070	0.752
5	1.000	0.138		2.04	0.100	0.710
6	1.000	0.200		3.06	0.070	0.650
7	1.000	0.200		3.06	0.100	0.810
8	1.000	0.270		4.08	0.070	0.830
9	1.000	0.270		4.08	0.100	0.802
10	1.000		0.891	1.02		1.784
11	1.000		1.510	1.53		2.451
12	1.0		2.103	2.04		3.063

A: 4,4'-methylenebis(phenylisocyanate); B: poly(1,4-butandiol)tolylene-2,4-diisocyanate terminated.

The same procedure was performed using ethandiol as reagent (Table 5). In this case lignin, 4,4'-methylenebis(phenylisocyanate), and ethandiol were suspended in THF for 5 h in the presence of catalytic amount of stannous octoate. After evaporation of the solvent the residue was treated in an oven at 72 °C overnight.

The reaction was carried out using poly(1,4-butandiol)tolylene-2,4-diisocyanate terminated (Table 5). In this case, lignin and the prepolymer were suspended in THF for 5 h in the presence of catalytic amount of stannous octoate. After solvent evaporation, the residue was maintained at 72 °C in an oven overnight (Table 5).

The synthesis of polyurethanes and polyesters involves the use of lignin as a source of hydroxyl groups. The total hydroxy groups in lignin was determined as described by Mansson [39] We obtained a value of 1.02 mEq g^{-1}. In the synthesis of polyurethanes we used two, three, and four equivalents of diisocyanate, respectively. The best result was obtained using two equivalents of diisocyanate.

The treatment of lignin with different amounts of 4,4'-methylenebis(phenylisocianate) gave the results reported in Table 6 (entries 1-3). After the treatment with the diisocyanate the obtained material was maintained at 72 °C in an oven to obtain crosslinking. The FTIR spectrum of this material showed the presence of a peak at 1730 cm^{-1} (urethane) and peaks at 1643, 1551, and 1237 cm^{-1} that could be attributed to the stretching of the carbonyl group in urea and to the N-H bending.

The ESEM of the same material (Figure 23) showed that there was no relation between the morphology of native lignin and the obtained polymer.

Polyurethanes were obtained also by using ethylene glycol as co-reagent (Table 5, entries 4-9). The presence of ethylene glycol reduced the yields of the polyurethanes. The best result was obtained when 0.270 g of the diisocyanate and 0.070 g of the glycol were used (Table 5, entry 8).

FTIR spectrum showed the same type of absorption described before (1730, 1643, 1547, and 1236 cm^{-1}). ESEM (Figure 24) showed that the morphological aspect of the new material was very different from that of lignin: while lignin showed the presence of grains of ca. 50 μm diameter, the polyurethane appeared as grains with dimension higher than 200 μm.

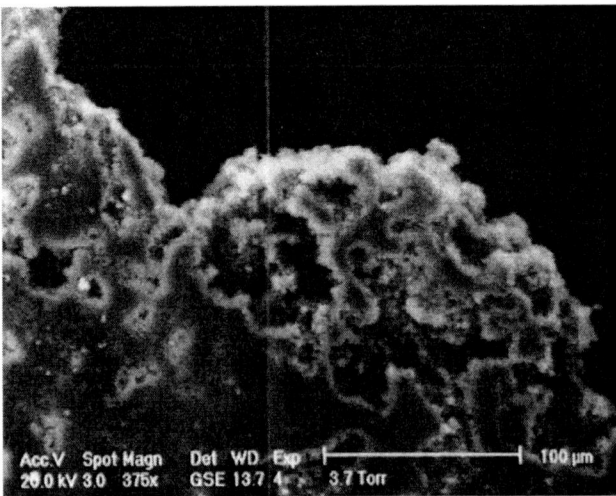

Figure 23. ESEM of the polyurethane obtained from lignin and 4,4'-methylenebis(phenylisocyanate).

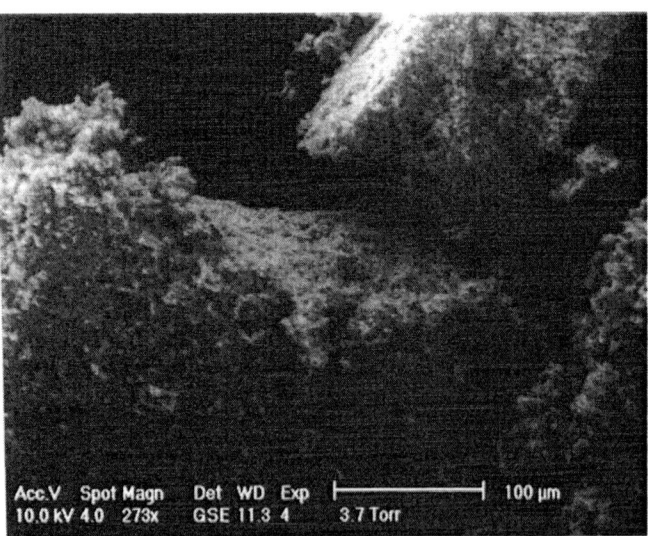

Figure 24. ESEM of the polyurethane from lignin and 4,4'-methylenebis(phenylisocyanate) in the presence of ethandiol.

We used also a poly(1,4-butandiol) terminated with tolylene-2,4-diisocynate prepolymer (Aldrich, M_n = 1600). The results of our experiments are reported in Table 5 (entries 10-12). We used 1, 1.5, and 2 equivalents of the prepolymer in relation to the hydroxy content of lignin. The use of this prepolymer gave the best results in the formation of polyurethanes.

While using 4,4'-methylenebis(phenylisocyanate) the obtained materials were not soluble in common solvents used for GPC analysis, the material obtained by using the butanediol prepolymer allowed this type of analysis. In Figure 25 we report the molecular weight distribution observed. In Table 6 we have collected the average molecular weights for all the polyurethanes thus obtained. We observed a large increase of the molecular weight in comparison with the lignin used for the experiments.

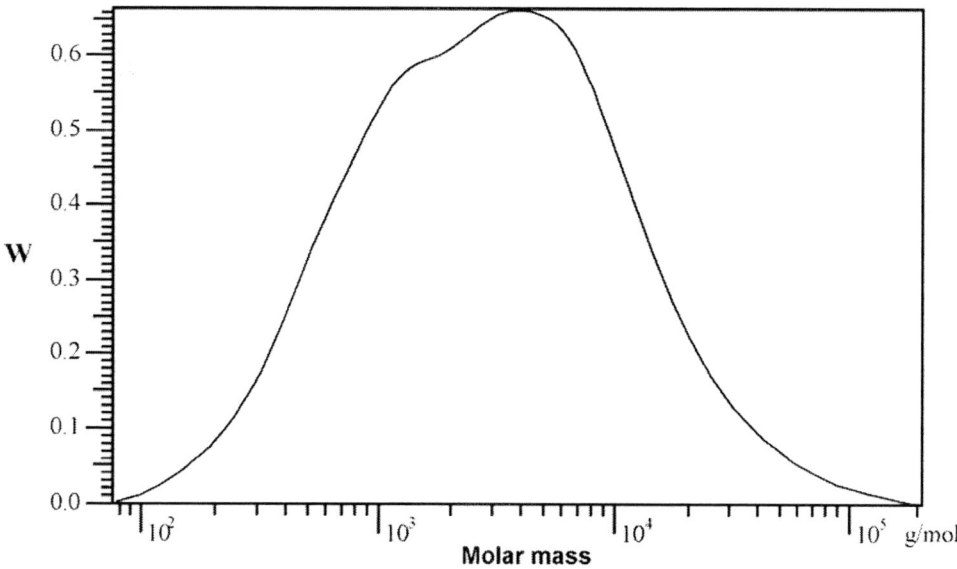

Figure 25. Molecular weight distribution in the polyurethane obtained from steam exploded lignin and poly(1,4-butandiol) terminated with tolylene-2,4-diisocyanate.

Table 6. Average molecular weights for polyurethanes obtained from steam exploded lignin from straw and poly(1,4-butandiol) terminated with tolylene-2,4-disocyanate

Entry	M_n	M_w	M_z	M_p
10	766	5517	35268	939
11	1301	7819	37927	3268
12	1371	6830	28857	4047

FTIR spectrum showed absorption at 1730, 1548, and 1223 cm^{-1}, in agreement with the formation of a polyurethanes (Table 7).

The ^{13}C NMR spectra of these materials showed peaks at δ 155, 154 (C=O), 148, 147 (C-4 in etherified guaiacyl units), 138.5, 138 (C-1 in etherified β-O-4 syringyl units), 137 (C-1 in syringyl units), 131 (C-2/C-6 in *p*-hydroxyphenyl units), 130, 126 (aromatic carbons), 116 (C-5 in etherified and non etherified guaiacyl units), 112 (C-2 in guaiacyl units), 107 (aromatic carbons), 104.5 (C-2/C-6 in syringyl units), 70, 64.5, 62.5, 61 (ethereal and alcoholic carbons), 56.5 (methoxy group), 33.5, 31.5, 30, 26, 24.5, 23, 18, 16, and 14.5 ppm. These data were in agreement with the presence of signals of both lignin and aliphatic ethereal prepolymer. The same conclusion could be obtained from ^{1}H NMR spectrum (Table 7).It showed signals due to the presence of lignin and prepolymer: in particular, the peaks at δ 1.3, 1.7, 2.0, 3.3 and 5.1 could be attributed to the presence of polybutandiol moiety. The ESEM image of the obtained materials showed the presence of very large grains (Figure 26). DSC analysis showed a T_g at -6 °C, assigned to the glass transition of the poly(1,4-butandiol) chains. In the temperature range used in DSC analysis no thermal decomposition was observed.

Table 7. FTIR and ^{1}H NMR data for polyurethane obtained from lignin and poly(1,3-butandiol)tolylene-2,4-diisocyanate terminated and for a polyester

Polyurethane		Polyester
IR absorptions [cm^{-1}]	^{1}H NMR peaks [δ, ppm]	IR absorption [cm^{-1}]
3357	8.71	3449
2941	7.54	2921
2857	7.22	2854
1730	5.45	1811
1602	5.04	1736
1542	4.28	1473
1450	3.80	1412
1370	3.32	1335
1227	2.36	1269
1111	2.08	1176
	1.60	1075

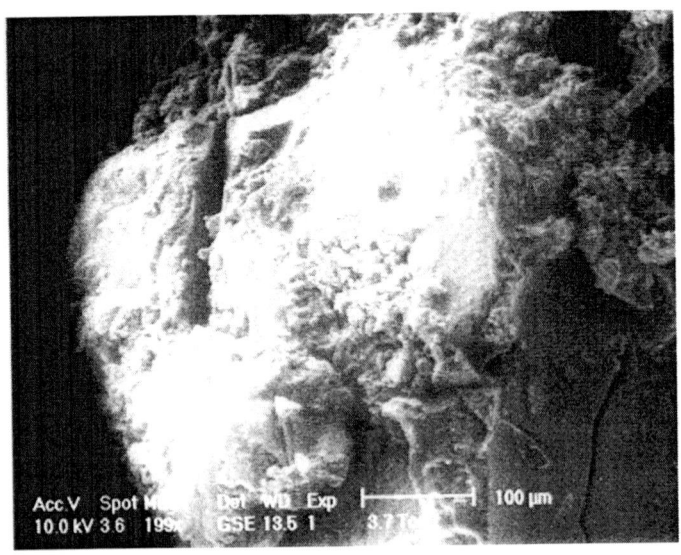

Figure 26. ESEM of the polyurethane obtained from steam exploded lignin and poly(1,4-butandiol) terminated with tolylene-2,4-diisocyanate.

SYNTHESIS OF POLYESTERS

We tested the possible use of this type of lignin in the preparation of polyesters. Lignin (1 g) and dodecandioyl dichloride were dissolved in THF (25 mL) in the presence of a stoichiometric amount (referred to the acyl chloride) of triethylamine. After 94 h, the mixture was extracted with ethyl acetate and dried over Na_2SO_4.

We used lignin from steam explosion as substrate in reactions with dodecandioyl dichloride. We used different lignin/dodecandioyl dichloride ratios on the basis of hydroxy content of lignin (Table 8).

Table 8. Polyesters from lignin

Entry	Lignin [g]	R(COCl)$_2$ [g]	mEq	Yield [g]	M_n	M_w	M_z
1	1.0	1.0	3.7	0.8	1915	18863	49371
2	1.0	0.7	2.6	0.6	2312	20153	355141
3	1.0	0.5	1.9	0.4	6382	29069	76416

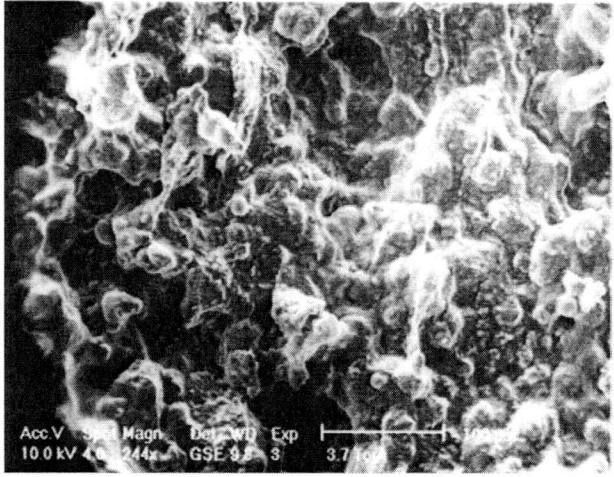

Figure 27. ESEM of the polymer between lignin and dodecandioyl dichloride.

In Table 8 we collected the average molecular weights of the obtained polymers. The polyester with the highest values of M_n, M_w, and M_z, and, then, with a low amount of molecules with high molecular weight, was obtained by using the conditions reported in entry 2. FTIR spectra were in agreement with the formation of polyesters (Table 7). The absorptions at 1736 (C=O stretching) and 1074 cm^{-1} (C-O stretching) were diagnostic for the presence of the ester function.

The ESEM image of the obtained polyester showed that the new material had a homogeneous structure (Figure 27).

^{13}C and ^{1}H NMR spectra of the polyesters showed signals in agreement with the presence of ester function . ^{13}C NMR spectrum showed signals at δ 179, 173.9, 169.6, 67.6, 67.4, 63.3, 44.4, 29.2, 29.1, 29.0, and 24.1 ppm. The ^{1}H NMR spectrum showed signals at δ 7.0, 4.1, 3.6, 2.4, 2.3, 1.8, 1.6, and 1.3 ppm.

CONCLUSION

Biomass has an enormous potential as a source of new interesting polymeric materials in a wide range of applications. Nevertheless, up to now only a small number of research groups is involved in such an activity. In the Basilicata region (Italy) agricultural residues are largely present, for example straw, which represents a particularly low-cost lignin source. Moreover, in the same region, Enea uses a particularly efficient and low environmental impact method (the Steam Explosion) of separation of lignin from cellulose and hemicellulose.

In this work we have used straw lignin, produced by Enea, blended with different commercial polymers, as polyethylenes and polystyrenes. The obtained blends are processable with the techniques in use for thermoplastics. Lignin stabilizes polystyrene and low-density polyethylene against the UV radiation. As far as the mechanical properties are concerned, the addition of lignin increases the modulus of blends, with respect to the neat polymers. Nevertheless, the presence of lignin decreases the tensile strength and the elongation at breaking. These results have been related with the poor compatibility between lignin and the synthetic apolar polymers and to the nonuniform distribution and poor adhesion of the lignin particles to the matrix. These problems may be overcome by using more efficient mixing techniques and a compatibilizing agent

On the basis of the results described above we have shown that steam explosion lignin from straw can be used as partner in wood adhesive: in fact it reacts thermally with diisocyanates to give polyurethanes. This result is particularly significant considering that in our experimental conditions only a gentle thermal treatment (very different from 90-170 °C used in the farm) was used. The presence of lignin as additive in wood adhesive could be important in order to have a more stable material [40,41]

In conclusion, we showed that lignin from straw obtained through steam explosion process can be used as starting material for the preparation of polyurethanes potentially useful in the formulation of wood adhesive and in the synthesis of polyesters, that can be used in the formulation of polyurethane coatings. Results on the possible use of lignin in the formulation of wood adhesive will be presented in the near future.

REFERENCES

[1]   Lin, S. Y.; Lebo, S. E. Jr.. In: *Encyclopedia of Chemical Technology*, Kirk-Othmer, New York, Wiley – Interscience 1995 (volume 4).

[2]   Glasser, W. G.; Sarkanen, S. *Lignin: Properties and Materials*. ACS Symposium series 398, Washington DC 1989, American Chemical Society.

[3]   Sanchez, C. G.; Exposito Alvarez L. A. Michromrchanics of lignin/polypropylene composites suitable for industrial applications, *Angew. Makromol. Chem.* 1999; *272* (1), 65-70

[4]   Alexy, P.; Kosikova, B.; Podstranka, G. The effect of blending lignin with polyethylene and polypropylene on physical properties. *Polymer* 2000, *41* (13), 4901-4908

[5]   Glasse W, Lignin. In: *Pulp and Paper*, 3rd Ed., Casey JP editor, New York, Interscience Publisher, 1981, p. 39

[6] Li, Y.; Mlynar, J.; Sarkanen, S. The first 85% kraft lignin-based thermoplastics. *J. Polym. Sci. Part B: Polym. Phys.* 1997, *35* (12), 1899-1910.

[7] Kosikova, B.; Demianova, V.; Kacurakova, M. Sulfur-free lignins as composites of polypropylene films. *J. Appl. Polym. Sci.*. 1993, *47* (6), 1065-1073

[8] Kosikova, B.; Revajova, A.; Demianova, V. The effect of adding lignin with polyethylene and polypropylene on physical properties. *Eur. Polym. J.* 1995, *31* (10), 953-956

[9] Kharade, A. Y.; Kale, D. D. Lignin-filled polyolefins. *J. Appl. Polym. Sci.* 1999, *72*(10), 1321-1326.

[10] Feldman, D.; Banu, D. Contribution to the study of rigid PVC polyblends with different lignins. *J. Appl. Polym. Sci.* 1997, *66*(9), 1731-1744

[11] Nitz, H.; Semke, H.; Mulhaupt, R. Influence of lignin type on the mechanical properties of lignin based compounds. *Macromol. Mater. Eng.* 2001, *286* (12), 737-743

[12] Cheradame, H.; Detoisien, M.; Gandini, A.; Pla, F.; Roux, G. Polyurethane from kraft lignin. *Br. Polym. J.* 1989, *21*, 269-275.

[13] Kelley, S. S.; Glasser, W. G.; Ward, T. C. Multiphase materials with lignin. 9. effect of lignin content on interpenetrating polymer network properties. *Polymer* 1989, *30*, 2265-2268.

[14] Yoshida, H.; Morck, R.; Kringstad, K. P.; Hatakeyama, H. Kraft lignin in polyurethanes. II, effectd of the molecular weight of kraft lignin on the properties of polyurethanes from a kraft lignin-polyethertriol-polymeric MDI system. *J. Appl. Polym. Sci.* 1990, *40*, 1819-1832.

[15] Reimann, A.; Morck, R.; Yoshida, H.; Hatakeyama, H.; Kringstad, K. P. Kraft lignin in polyurethanes. III. Effects of molecular weight of PEG on the properties of polyurethanes from kraft lignin-PEG-MDI system. *J. Appl. Polym. Sci.* 1990, *41*, 39-50.

[16] Kelley, S. S.; Ward. T. C.; Glasser, W. G. Multiphase materials with lignin. VIII. Interpenetrating polymer networks from polyurethanes and poly(methyl methacrylate). *J. Appl. Polym. Sci.* 1990, *41*, 2813-2828.

[17] Thring, R. W.; Vanderlaan, M. N.; Griffin, S. L. Polyurethanes from Alcell lignin. *Biomass Bioenergy* 1997, *13*, 125-132.

[18] Vanderlaan, M. N.; Thring, R. W. Polyurethanes from Alcell lignin fractions obtained by sequential solvent extraction. *Biomass Bioenergy* 1998, *14*, 525-531.

[19] Evtuguin, D. V.; Andreolety, J. P.; Gandini, A. Polyurethanes based on oxygen-organosolv lignin. *Eur. Polym. J.* 1998, *34*, 1163-1169.

[20] Sarkar, S.; Adhikari, B. Synthesis and characterization of lignin-HTPB copolyurethane. *Eur. Polym. J.* 2001, *37*, 1391-1401.

[21] Kundu, S. K.; Ray, P. K.; Day, A.; Sen, S. K. Infrared spectra of acrylonitrile-grafted jute fibers. *J. Appl. Polym. Sci.* 1989, *38*, 1951-1955.

[22] Lathia, A.; Chang, F. F.; Meister, J. J. Effect of class, order, family, genus, species, and recovery method of lignin on product properties of grafted lignin. *Polym. Prep.* 1990, *31*, 648-649.

[23] Meister, J. J.; Li, C. T. Synthesis of cationic graft copolymers of lignin. *Polym. Prep.* 1990, *31*, 653-654.

[24] Lathia, A.; Meister. J. J. Formation of lignin-alkoxy polyols from yellow poplar lignin. *Polym. Prep.* 1990, *31*, 660-661.

[25] Meister, J. J.; Lathia, A.; Chang, F. F. Solvent effects, species and extraction method effects, and coinitiator effects in the grafting of lignin. *J. Polym. Sci., Part. A: Polym. Chem.* 1991, *29*, 1465-1473.

[26] Meister, J. J.; Li, C. T. Graft 1-phenylethylene copolymers of lignin.1. Synthesis and proof of copolymerization. *Macromolecules* 1992, *25*, 611-616.

[27] Gunnels, D. W.; Gardner, D. J.; Chen, M. J.; Meister, J. J. Alteration of surface energy of wood using thermoplastic graft copolymers. *Polym. Mater. Sci. Eng.* 1992, *67*, 227.

[28] Meister, J. J.; Zhao, Z. Lignin graft copolymers containing a methyl methacrylate graft side chain. *Polym. Mater. Sci. Eng.* 1992, *67*, 228-229.

[29] Meister, J. J.; Chen, M. J. J Graft copolymers of wood pulp and 1-phenylethylene. I. Generality of synthesis and proof of copolymerization. *Appl. Polym. Sci.* 1993, *49*, 935-951.

[30] Chen, M. J.; Meister, J. J.; Gunnels, D. W.; Gardner, D. J. Binding to a hydrophobic surface by altering of the surface energy of wood using graft copolymers. *Polym. Mater. Sci. Eng.* 1993, *68*, 243-244.

[31] Meister, J. J.; Zhao, Z. Physical characterization of lignin graft copolymers with poly(methyl methacrylate) sidechains. *Polym. Prep.* 1993, *34*, 606-607.

[32] Meister, J. J.; Aranha, A.; Wang, A. Poly(3-hydroxybutyrate)-3-(hydroxyvalerate)-lignin graft copolymer blends. *Polym. Prep.* 1993, *34*, 608-609.

[33] Gandini, A.; Naceur, B. M.; Guo, Z. X.; Montanari, S. In Chemical modification, properties and usage of lignin, Hu, T. Q., Ed.; Kluwer Academic/Plenum Publisher: New York, 2002, p. 57.

[34] Glasser, W. G.; Jain, R. K. In Chemicals and materials from renewable resources, Bozell, J. J., Ed.; American Chemical Society: Wasshington DC, 2001, p 191.

[35] Evtugin, D. V.; Gandini, A. Polyesters based on oxygen-organosolv lignin *Acta Polym.* 1996, 47, 344-350.

[36] Guo, Z. X.; Gandini, A.; Pla, F. Polyesters from lignin. 1. The reaction of kraft lignin with dicarboxylic acid chlorides. *Polym. Int.* 1992, 27, 17-22.

[37] Guo, Z. X.; Gandini, A. Polyesters from lignin. 2. The copolyesterification of kraft lignin and polyethylene glycols with dicarboxylic acid chlorides. *Eur. Polym. J.* 1991, *27*, 1177-1180.

[38] Mc Crum, N.C.; Read, B. E.; Williams, G. Anelastic and Dielectric Effects in Polymeric Solids, New York, Wiley-Interscience 1967.

[39] Chen, C. –L. In Methods in lignin chemistry, Lin, S. Y.; Dence, C. W., Eds; Springer-Verlag: Berlin, 1992, Chap 7.1.

[40] Ferri, R. Uses of lignin for an environmental sustanaible development. *EPA Newsletter* 2004, *77*, 50-53.

[41] Pucciariello, R.; Villani, V.; Bonini, C.; D'Auria, M.; Vetere, T. Physical properties of straw lignin-based polymer blends. *Polymer* 2004, *45*, 4159-4169.

INDEX

D

experimental design, 33, 35, 36, 42, 51
exploitation, 13, 138
exposure, 20
external benefits, 125
external costs, 58
extinction, 58
extraction, 10, 11, 15, 16, 17, 18, 19, 21, 28, 30, 31,
 60, 73, 74, 75, 151, 172, 203, 204
extrusion, 24, 180, 183

F

fabric, 134
factor analysis, 49
failure, 134, 184
family, 42, 67, 105, 147, 203
farmers, 119, 121
farms, 109
fat, 7, 15, 21, 143, 153, 173
fatty acids, 4, 15, 142, 143, 149, 159, 172
fermentation, 9, 10, 12, 14, 17, 18, 24, 25, 29, 30, 94
fertilization, 149
fertilizers, 95
fibers, 178, 203
filament, 12
fillers, 196
film formation, 20
films, 22, 24, 179, 203
filtration, 14, 17, 18, 29, 159
finance, 99, 137, 138
financial support, 96, 139
financing, 126, 127, 139
Finland, 50, 91
firms, 116, 117
First World, 50, 51
fishing, 127
fission, 2
fixed costs, 130
flame, 22
flavor, 21
flexibility, 20, 59, 77, 93, 111, 133, 182
fluctuations, 14, 130
flue gas, 40, 42, 43, 56, 62, 66, 75
fluid, 8
fluidized bed, 51
foams, 12, 20, 21
focusing, 134
foils, 24
food, vii, 1, 2, 5, 10, 11, 15, 16, 18, 19, 20, 22, 23,
 24, 26, 27, 30, 31, 34, 63, 129, 137, 142, 157, 158
food industry, 12, 24
food products, 22, 158
Ford, 162

forests, viii, 53, 57, 118, 134
fossil, viii, 6, 34, 53, 54, 55, 57, 58, 60, 61, 64, 66,
 73, 75, 91, 93, 95, 96, 97, 103, 104, 132, 139
fractal, 10
France, 27, 148
freezing, 17, 28
frequency distribution, 85, 88
friction, 183
FTIR, 197, 199, 200, 201
fuel, vii, viii, ix, x, 34, 35, 45, 50, 55, 58, 60, 61, 62,
 67, 69, 70, 71, 73, 74, 75, 84, 91, 92, 93, 94, 95,
 98, 101, 103, 104, 105, 106, 118, 119, 120, 121,
 123, 125, 126, 129, 130, 131, 132, 134, 136, 138,
 139, 141, 142, 143, 145, 146, 147, 152, 153, 154,
 155, 156, 157, 158, 161, 162, 163, 165, 166, 167,
 169, 171, 172, 173, 174, 175, 178
fuel efficiency, 75, 123
fuels, 2
fulfillment, 135
funding, 113, 137
funds, 111, 113, 118, 127, 137
fusion, 2
future, 2, 14, 16, 27

G

garbage, vii
gasification, viii, ix, 3, 51, 54, 60, 62, 67, 69, 71, 76,
 92, 93, 94, 98, 100, 101, 102, 106, 108, 112, 128,
 129, 132, 133, 136, 137, 138
GDP, 82
gel, xi, 14, 18, 177
gel permeation chromatography, xi, 177
gelation, 20, 21
gender, 137
generation, ix, 34, 46, 56, 58, 61, 69, 92, 101, 102,
 103, 104, 105, 106, 107, 109, 111, 113, 114, 115,
 118, 119, 120, 121, 122, 123, 126, 127, 128, 130,
 131, 132, 134, 135, 136, 137, 138, 139, 140, 142
genotype, 16
Germany, 1, 6, 8, 26, 28, 50
glass transition, xi, 177, 180, 182, 185, 186, 187,
 189, 195, 199
glass transition temperature, 182, 185, 186, 187, 195
glassy polymers, 185, 191
global trade, 84
glutamic, 2, 4
glutamic acid, 2, 4
glycerin, ix, 141, 143, 144
glycerol, 160, 163
glycol, 24, 197
glycoproteins, 14
goals, viii, 101